W0091363

Computer-Assisted Reservoir Management

Computer-Assisted Reservoir Management

Editor

Gaurav Vashishth

Computer-Assisted Reservoir Management

Edited by **Gaurav Vashishth**

Printed in 2017

ISBN: 978-1-68117-346-7

Library of Congress Control Number: 2015939259

© 2016 by

SCITUS Academics LLC,
616, Corporate Way, Suite 2, 4766,
Valley Cottage, NY 10989

www.scitusacademics.com

This book contains information obtained from highly regarded resources. Copyright for individual articles remains with the authors as indicated. All chapters are distributed under the terms of the Creative Commons Attribution License, which permits unrestricted use, distribution, and reproduction in any medium, provided the original author and source are credited.

Notice

Reasonable efforts have been made to publish reliable data and views articulated in the chapters are those of the individual contributors, and not necessarily those of the editors or publishers. Editors or publishers are not responsible for the accuracy of the information in the published chapters or consequences of their use. The publisher believes no responsibility for any damage or grievance to the persons or property arising out of the use of any materials, instructions, methods or thoughts in the book. The editors and the publisher have attempted to trace the copyright holders of all material reproduced in this publication and apologize to copyright holders if permission has not been obtained. If any copyright holder has not been acknowledged, please write to us so we may rectify.

Contents

Preface

Computer-assisted- is a prefix that hints to the use of a computer as an indispensable tool in a certain field, usually derived from more traditional fields of science and engineering. Reservoir management is a dynamic process that recognizes the uncertainties in reservoir performance resulting from our inability to fully characterize reservoirs and flow processes. It seeks to mitigate the effects of these uncertainties by optimizing reservoir performance through a systematic application of integrated, multidisciplinary technologies. It approaches reservoir operation and control as a system, rather than as a set of disconnected functions. As such, it is a strategy for applying multiple technologies in an optimal way to achieve synergy.

Editor

Developing a Robust Surrogate Model of Chemical Flooding Based on the Artificial Neural Network for Enhanced Oil Recovery Implications

Mohammad Ali Ahmadi

Department of Petroleum Engineering, Ahwaz Faculty of Petroleum Engineering, Petroleum University of Technology, Ahwaz 6198144471, Iran

ABSTRACT

Application of chemical flooding in petroleum reservoirs turns into hot topic of the recent researches. Development strategies of the aforementioned technique are more robust and precise when we consider both economical points of view (net present value, NPV) and technical points of view (recovery factor, RF). In current study many attempts have been made to propose predictive model for estimation of efficiency of chemical flooding in oil reservoirs. To gain this end, a couple of swarm intelligence and artificial neural network (ANN) is employed. Also, lucrative and high precise chemical flooding data banks reported in previous attentions are utilized to test and validate proposed intelligent model. According to the mean square error (MSE), correlation coefficient, and average absolute relative deviation, the suggested swarm approach has acceptable reliability, integrity and robustness. Thus, the proposed intelligent model can be considered as an alternative model to predict the efficiency of chemical flooding in oil reservoir when the required experimental data are not available or accessible.

INTRODUCTION

The oil and gas upstream industries are recently encountered with the difficulties and challenges of dealing with hydrocarbon resources whose productions with conventional technologies are following an upward trend of technical limitations. It is because of achieving the stage of decline phase by most of oilfields around the world. Therefore, how to postpone the abandonment of reservoirs has tuned into the priority of researchers in the worldwide. Their researches normally highlight the concept of great necessities for inventions of new techniques, normally classified as tertiary oil recovery methods, having abilities of maintaining the economic production rate [1–3].

Chemical enhanced oil recovery approaches as one of the most effective subsets of tertiary methods are known as a key to unlock the exploitation of referred resources. Different methods for this process have been developed, such as polymer, surfactant/polymer (SP), and alkaline/surfactant/polymer (ASP) flooding. These methods are applied to increase the rate of oil production through focusing on both lowering the interfacial tension and reducing the water mobility. In more details, it has enormously been declared in previous literatures that in order to design, manage, and run a chemical enhanced oil recovery operation it is highly required to set very expensive and time-consuming but precise experimental procedures which their generated results must be gained to plan effectively the process of injecting chemical materials [4–9].

The laboratorial generated outputs are then used to conclude two parameters, recovery factor (RF) and net present value (NPV), which are used to evaluate the performance of the chemical flooding which is one of the most popular methods of chemical enhanced oil recovery. Having knowledge about these two parameters is essentially vital to make decisions if it is beneficial to run the referred operation. Unfortunately, there are no global methods to interpret simultaneously both aforementioned factors although there are numerous numbers of different software and numerical or analytical methods which are capable of making very precise quantitative decisions about the amount of one of the RF or NPV [10–12].

Hence, there is a great need in oilfield for having access to a solution or model which can predict the amount of these two parameters at the same time. The major aim of current study is to execute new kind of artificial intelligence approaches to suggest robust and accurate predictive method to forecast efficiency of the chemical flooding through petroleum reservoirs. To gain successfully this referred goal, hybridization of artificial neural network and particle swarm optimization (PSO) was executed on the previous literature data bases. The integrity and performance of the proposed predictive approaches in estimating recovery factor (RF) and net present value (NPV) from the literature are described in details.

DATA GATHERING

The data utilized throughout this research have been gathered from previous attentions [9] in which chemical flooding had been simulated in Benoist sand reservoir, by executing UTCHEM simulator. That reservoir has been produced under primary and secondary processes over fifty years. The original dataset contained 202 data. Each data had 7 inputs: surfactant slug size, surfactant concentration in surfactant slug, polymer concentration in surfactant slug, polymer drive size, and polymer concentration in polymer drive, K_v/K_h ratio, and salinity of polymer drive. In addition, the outputs were RF and NPV. The ranges of implemented data banks are reported in Table 1 [9].

Table 1: Statistical analysis of the implemented chemical flooding data samples [9]

Parameter	Unit	Type	Min.	Max.	Average	Standard deviation
Surfactant slug size	PV	Input	0.097	0.259	0.177	0.072
Surfactant concentration	Vol. fraction	Input	0.005	0.03	0.017	0.011
Polymer concentration in surfactant slug	wt.%	Input	0.1	0.25	0.177	0.067
Polymer drive size	PV	Input	0.324	0.648	0.482	0.144
Polymer concentration in polymer drive	wt.%	Input	0.1	0.2	0.148	0.044
V/ h ratio	—	Input	0.01	0.25	0.129	0.107
Salinity of polymer drive	Meq/mL	Input	0.3	0.4	0.349	0.045
Recovery factor (RF)	%	Output	14.82	56.99	39.67	9.24
Net present value (NPV)	$ MM	Output	1.781	7.229	4.45	1.53

ARTIFICIAL NEURAL NETWORK AND PARTICLE SWARM OPTIMIZATION

Artificial neural network (ANN) includes simple nodes, named as neurons, which are bonded to each other to construct a network model. Indeed, the biological nervous systems can be simulated with the ANN system, somehow. Characterization of an ANN model is normally performed through three ways including (a) certain patterns between various layers, (b) connection between input and output via activation function, and (c) updating the interconnection weights through training process [13–24].

In fact, the main purpose of an ANN model is to determine target function through internal computation during the training phase if the values of input variables are provided. The most common type of ANN is the multilayer feed forward neural network which is made up of group of interconnected neurons organized in the form of layers: input layer, hidden layer(s), and output layer where each layer comprises a group of neurons as presented in Figure 1. This network is strictly an acyclic type since signals propagate only in a forward direction from the input neurons to the output neurons and no signals are allowed to be fed-back among the neurons. The number of neurons in the input and output layers is decided by the number of input and output variables that are planned for the predictive tool. However, the optimal number of neurons in hidden layer(s) is a strong function of nonlinearity and dimensionality of the problem under study [13–24].

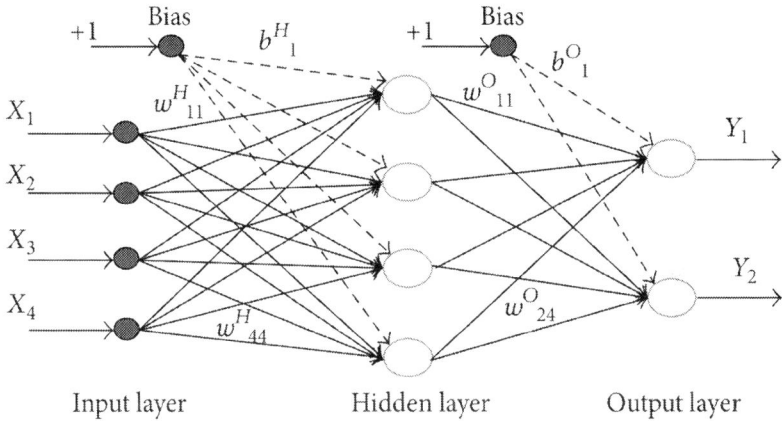

Figure 1: Architecture of multilayer feed forward neural network. The symbol w^H_{qr} denotes the synaptic weight between the output of the rth neuron in the hidden layer and the input of the qth neuron in output layer. The symbol b^H_q denotes the bias of the qth neuron in hidden layer. The superscript O stands for output layer.

The artificial neuron is the fundamental part of the neural networks. Each artificial neuron—excluding neurons at the input layer—takes and processes inputs gathered from other neurons. Given further information, each artificial neuron is a mathematical information-processing unit. The processed information is presented at the output end of the neuron. Figure 2 addresses the procedure in which an artificial neuron treats the data and information entered in the model. Each input signal (a_k) is primarily multiplied by the corresponding weight value (w_{kj}) and the resultant products are summed up to generate a total weight in the form of $w_{j1}a_1 + w_{j2}a_2 + \cdots + w_{jm}a_m$. The sum of the weighted inputs and the bias $(S_j = \sum_{k=1}^{m} w_{jk} \cdot a_k + b_j)$ forms the input to the activation function, φ. An activation function processes this sum and gives out the output, oj. Indeed, the resulting sum is processed by a neuron activation function to obtain the ultimate output of the neuron as follows [13–26]:

$$o_j = \varphi\left(S_j\right) = \varphi\left(\sum_{k=1}^{m} w_{jk} \cdot a_k + b_j \right).$$

(1)

This output will be the input signal for the neurons in the following layer. The linear (purelin) transfer, tan-sigmoid (tansig) activation, and log-sigmoid (logsig) activation functions are mostly employed in the practical cases with applications in science and engineering disciplines. The corresponding relationships for these functions are defined, respectively, by (2)–(4), as given below [13–27]:

$$\varphi\left(s\right) = s,$$

(2)

$$\varphi\left(s\right) = \frac{e^s - e^{-s}}{e^s + e^{-s}},$$

(3)

$$\varphi\left(s\right) = \frac{1}{1 + e^{-s}}.$$

(4)

The weight factors are generally considered as the adaptive parameters in the network to obtain the strength of the input signals. A bias is characterized with a weight which is not responsible for connecting an input of two neurons to an output. A particular level of a neuron output signal is represented by a set of bias that does not depend on the input signals. The weight factors and biases are tuned during the course of training phase such that the network is able to forecast the accurate target parameter for a given set of inputs. There are a number of training algorithms with different methodologies in the context of intelligence system. A variety of optimization tools such as particle swarm optimization (PSO) [15, 18, 19], genetic algorithm (GA) [21], hybrid genetic algorithm and particle swarm optimization (HGAPSO) [13, 16], unified particle swarm optimization (UPSO) [14], and imperialist competitive algorithm (ICA) [17, 20, 23] for weight training of neural networks have been used.

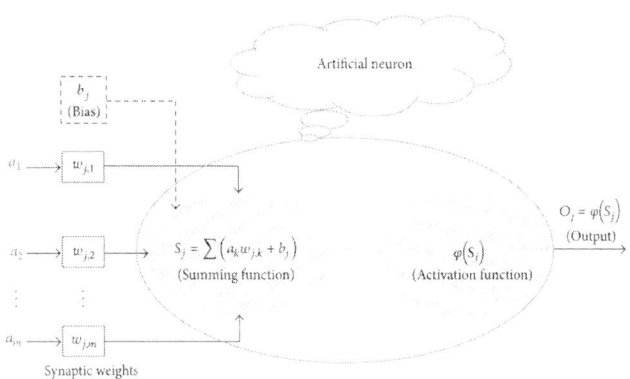

Figure 2: Information processing by an artificial neuron.

Kennedy [27] introduced the PSO as a strong stochastic optimization technique which simulates the social manners of birds within a group, based on population concept. It searches for an optimum solution by iteratively updating a swarm of particles.

The model originally includes a group of random particles (solutions). A random velocity is attributed to each candidate particle which flies within the problem space. The solutions consist of memory and try to attain the best position or/and fitness. This parameter is symbolized by "P_{best}" that is linked only to a specific particle. The model also retains the best fitness, known as "g_{best}," which is found among the entire solutions (particles) in the swarm. The candidate particle that obtains this fitness is the global best in the population [25–28]. In the current study, a particle's fitness is calculated through determination of the network output for every point in the training part and then computing the sum of squares of the resultant errors (MSE) for performance evaluation. The basic PSO theory involves variation of each particle velocity toward its P_{best} and g_{best} locations at each time interval. The particles' new velocity and position are updated according to the following equations [13–28]:

$$v_i^{n+1} = \omega v_i^n + c_1 r_1^n \left[x_{i,P_{best}}^n - x_i^n \right]$$

$$\tag{5}$$

$$+ c_2 r_2^n \left[x_{g_{best}}^n - x_i^n \right],$$

$$x_i^{n+1} = x_i^n + v_i^{n+1},$$

$$\tag{6}$$

where V_i^n and V_i^{n+1} are velocities of particle i at iterations n and $n+1$; $xi\,n$ and x_i^{n+1} are positions of particle i at iterations n and $n+1$; ω represents the inertia weight that directs the exploitation and exploration of the search space as it continuously updates velocity; $c1$ and $c2$ are termed as cognition and social components, respectively. They are considered as the acceleration constants which alter the velocity of a solution in the direction of p_{best} and g_{best} [13–28]; and r_1^n and r_2^n refer to the two random variables uniformly distributed in the interval of [0, 1].

Herein, PSO algorithm has been used in evolving weights of multilayer feed forward neural network. In this case, a particle's position at any iteration is described as a particle whose coordinates are connection weights. The vectors of weights for each particle i will be called x_i. Throughout the training process the above equations (equations (5) and (6)) will customize the network weights until a criterion is met. In this case, a lower MSE, as a sufficiently good fitness, is achieved; nevertheless, a maximum number of iterations are used to terminate the iterative search process if no improvement is observed over a number of consecutive generations in an appropriate time. The flowchart of the PSO-based training algorithm for the ANN is shown in Figure 3.

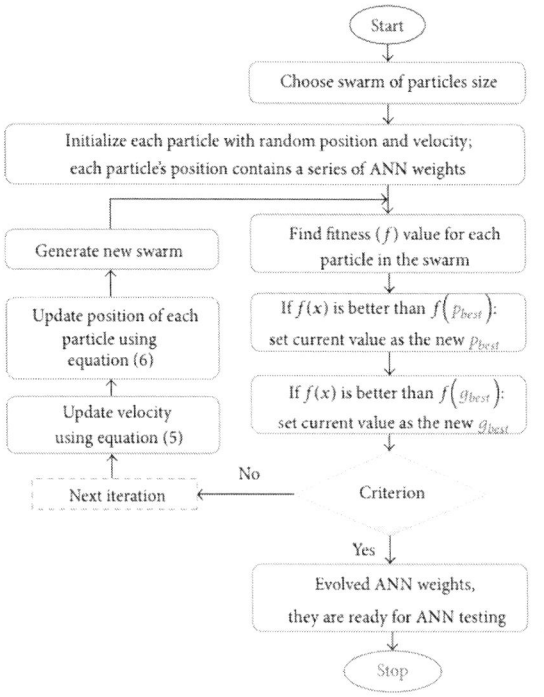

Figure 3: PSO-based algorithm flowchart in optimization of the weights of ANN.

The PSO utilizes a random procedure in the search space of the problem such that particles in the population are directed toward optimum positions but not in or between optimal areas [27]. Thus, PSO can be used to train neural networks with nondifferentiable (even discontinuous) neurons activation functions. It can be also implemented in cases where gradient or error information is not accessible. PSO is easy to implement and there are few parameters to be adjusted. However, the uniqueness of the algorithm lies in the dynamic interactions among the particles that turn it into a social-psychological model of knowledge management [27].

RESULTS AND DISCUSSION

According to the study accomplished by Cybenko [29], a network that consists of only one single hidden layer has the ability to approximate nearly any kind of nonlinear function. However, determination of the ideal number of neurons in the hidden layer is a challenging task; few neurons will not give adequate precision and too many hidden neurons may lead to overfitting. It means that the training data might be fitted adequately; however considerable oscillations between the points are noticed in the fitting curve, resulting in poor interpolation and extrapolation. The network performance is evaluated as demonstrated in Figure 4, when different number of neurons is tested. A smart model with one hidden layer (including just one neuron) was primarily built in the current study to predict the recovery factor and net present value (NPV) of chemical flooding in oil reservoirs. Prediction accuracy was further analyzed by an increase in the number of neurons to 10 to decide on the most precise technique. As clear from the results demonstrated in Figure 4, a 3-6-1 architecture (6 neurons in the hidden layer, 3 neurons in the input layer, and one neuron in output layer) offers the best model for recovery factor and net present value (NPV) prediction in terms of MSE and R^2, since the optimum structure achieved by the trial and error procedure has a very low mean squared error of MSE=0.0012 and a satisfactory coefficient of determination of R^2=0.9996, on the basis of comparison between the predicted and real data.

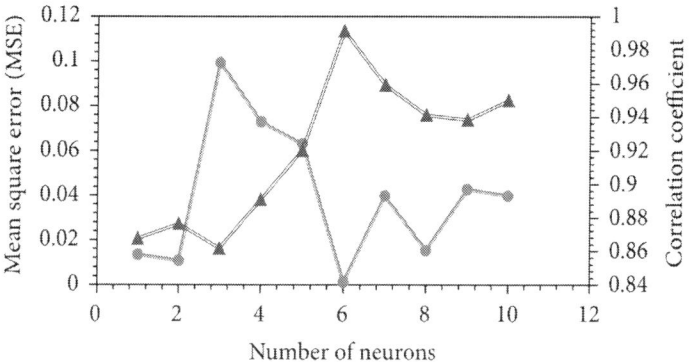

(a)

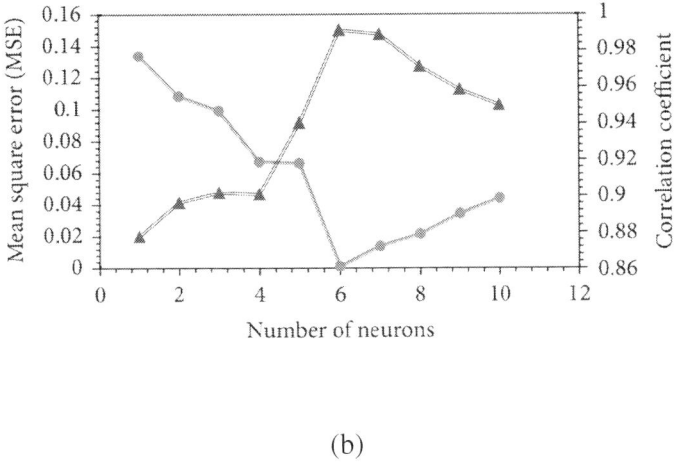

(b)

Figure 4: Effect of number of hidden neuron on PSO-ANN accuracy of (a) recovery and (b) NPV predictions in terms of MSE and -squared.

The generated results of the proposed intelligent approach are depicted through Figures 5 to 10. The existing contrasts between suggested intelligent approach and related recovery factor (RF) of the chemical flooding in oil reservoir in the regression plot have been depicted in Figure 5. As shown in Figure 5 which is a graphical and scatter presentation of the PSO-ANN results versus corresponding determined recovery factor (RF) data, the PSO-ANN outputs lie over the line Y=X, the fact that indicates the identity of outputs gained from suggested PSO-ANN model and relevant recovery factor data samples. To serve better understanding about generated results of the proposed PSO-ANN model, the comparison between gained recovery factor from the addressed model and real recovery factor data versus relevant data index has been illustrated in Figure 6. As illustrated in Figure 6, the obtained results of proposed model are as close as possible to real recovery factor (RF) data samples. To put it another way, the outputs of the PSO-ANN approach have the same behaviour as actual data do. The high considerable level of efficiency and accuracy related to the PSO-ANN approach in prediction of the recovery factor dataset of chemical flooding has once again been certified in Figure 6. Moreover, the robustness of the PSO-ANN has been demonstrated in terms of the relative deviations of PSO-ANN model outputs from corresponding determined recovery factor data in Figure 7. As could be observed in

Figure 7, the highest deviations of the suggested approach results are subjected to the early boundary of recovery factor data samples. 5% is the maximum degree of relative deviation shown in Figure 7.

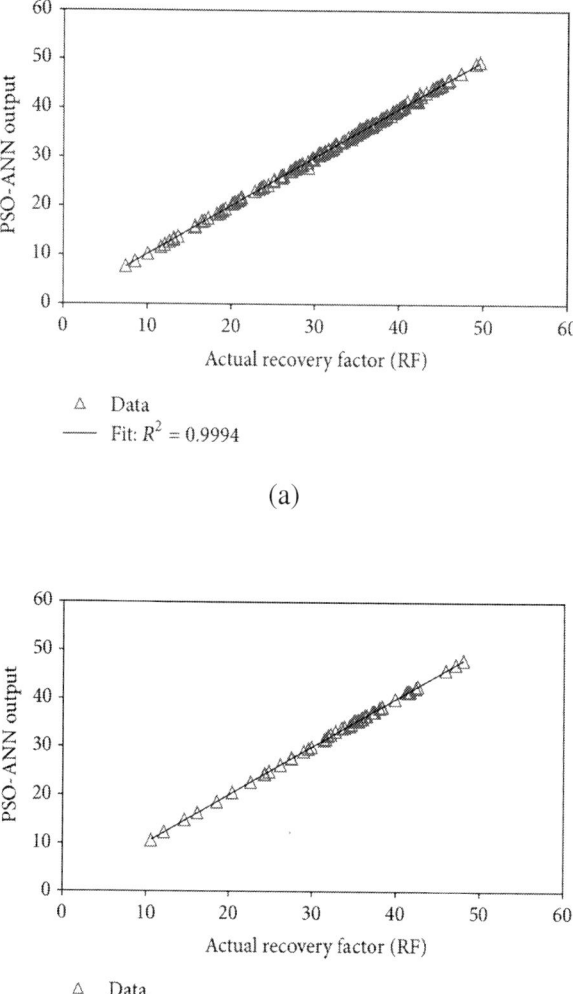

(a)

(b)

Figure 5: Performance plot of the suggested network model for determining recovery factor of chemical flooding owing to correlation coefficient (R^2): (a) training phase and (b) testing phase.

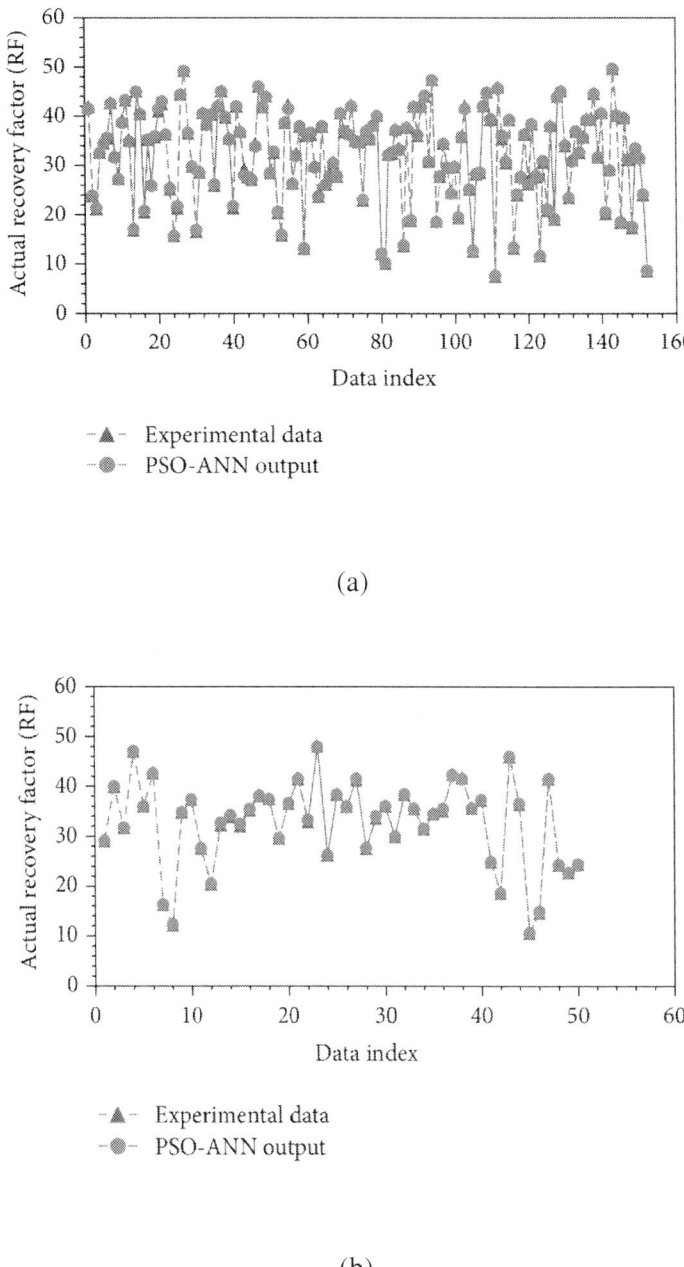

(a)

(b)

Figure 6: Comparison between suggested network model and recovery factor versus relevant data index: (a) training phase and (b) testing phase.

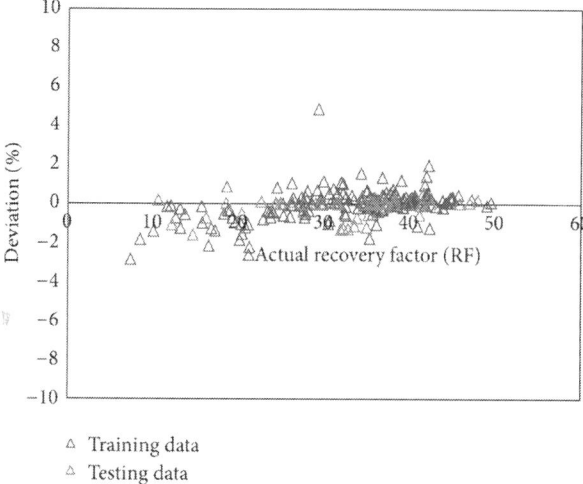

Figure 7: Relative error distribution of the proposed approach versus actual recovery factor (RF).

The draw parallel between our proposed intelligent PSO-ANN model results and related net present value (NPV) of the chemical flooding in oil reservoir in the regression plot has been shown in Figure 8. As shown in Figure 8 which is a graphical and scatter presentation of the PSO-ANN results versus corresponding determined net present value (NPV) data, the PSO-ANN outputs lie over the line Y=X, the fact that indicates the identity of outputs gained from suggested PSO-ANN model and relevant net present value (NPV) data samples. The comparison between generated net present value (NPV) from the addressed approach and real net present value (NPV) data versus relevant data index has been shown in Figure 9. As illustrated in Figure 9, the obtained results of proposed model are as close as possible to net present value (NPV) data samples. To put it another way, the outputs of the PSO-ANN approach have the same behaviour as actual data do. Furthermore, the effectiveness of the proposed intelligent model has been depicted in terms of the relative deviations of PSO-ANN model outputs from corresponding indicated net present value (NPV) data in Figure 10. As can be seen from Figure 10, the highest deviations of the suggested approach results are subjected to the early boundary of net present value (NPV) data. 6% is the maximum degree of relative deviation depicted in Figure 10.

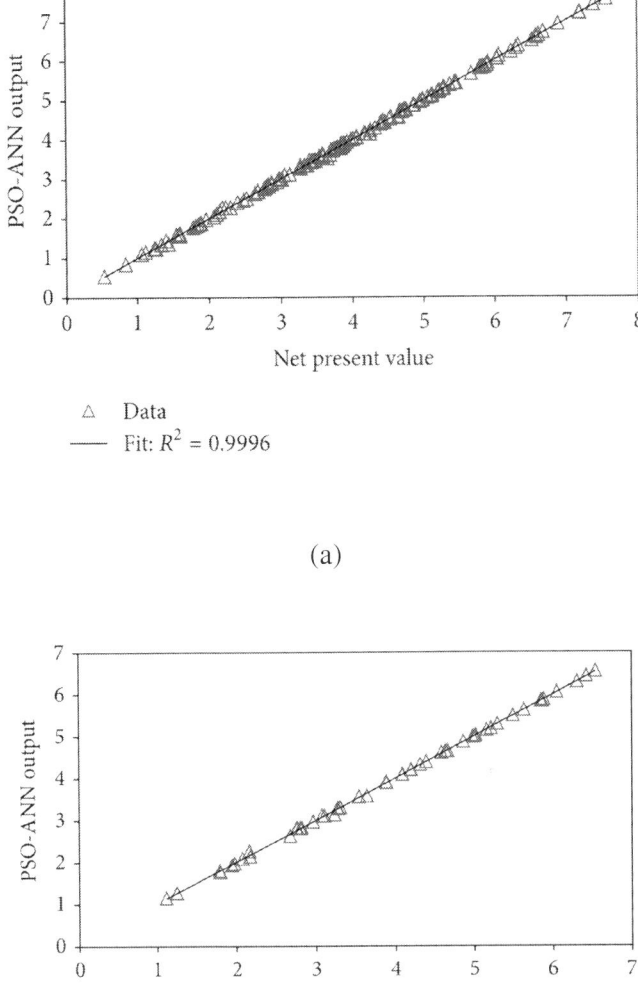

(a)

(b)

Figure 8: Performance plot of the suggested network model for determining net present value (NPV) of chemical flooding owing to correlation coefficient (R^2): (a) training phase and (b) testing phase.

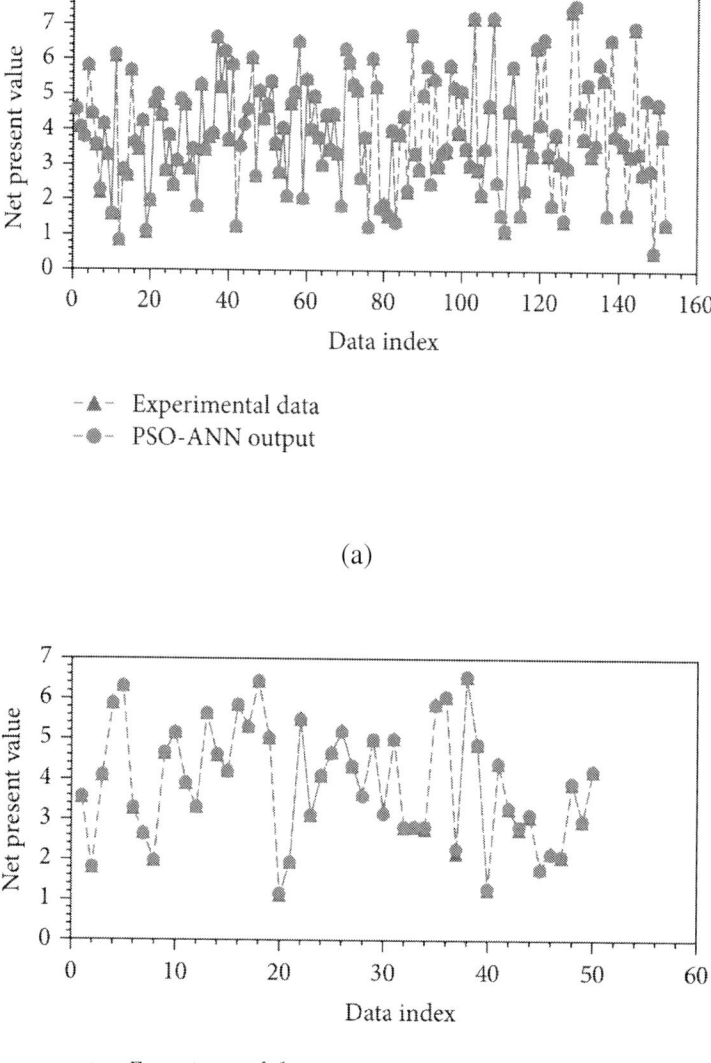

(a)

(b)

Figure 9: Comparison between suggested network model and net present value (NPV) versus relevant data index: (a) training phase and (b) testing phase.

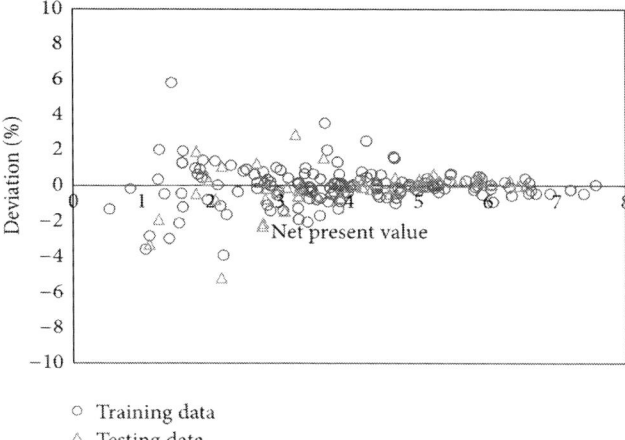

Figure 10: Relative error distribution of the proposed approach versus actual net present value (NPV).

The performance efficiency of the selected network is assessed using the various error analysis parameters. Table 2 tabulates the PSO-ANN accuracy in terms of correlation coefficient (R), coefficient of determination (R^2), mean absolute error (MAE), and mean squared error (MSE), which are defined as follows:

$$R = \frac{\sum_{i=1}^{N}\left(y^{T}_{i} - \overline{y^{T}}\right)\left(y^{P}_{i} - \overline{y^{P}}\right)}{\sqrt{\sum_{i=1}^{N}\left(y^{T}_{i} - \overline{y^{T}}\right)^{2}\sum_{i=1}^{N}\left(y^{P}_{i} - \overline{y^{P}}\right)^{2}}},$$

$$R^2 = 1 - \frac{\sum_{i=1}^{N}\left(y^{T}_{i} - y^{P}_{i}\right)^{2}}{\sum_{i=1}^{N}\left(y^{T}_{i} - \overline{y^{T}}\right)^{2}},$$

$$\text{MAE} = \frac{1}{N}\sum_{i=1}^{N}\left|y^{T}_{i} - y^{P}_{i}\right|,$$

$$\text{MSE} = \frac{1}{N}\sum_{i=1}^{N}\left(y^{T}_{i} - y^{P}_{i}\right)^{2},$$

$$(7)$$

in which N represents the total number of data points including either training, testing, or whole data set (input and output pairs), y^T_i refers to the actual value at the sampling point i, y^p_i is the ith output of the model, and $\overline{Y^T}$ and $\overline{Y^p}$ stand for the average magnitudes of the actual and predicted data, respectively.

Table 2: Statistical parameters of the proposed approaches in prediction of efficiency of chemical flooding in oil reservoirs

PSO-ANN		
	RF	NPV
Correlation coefficient (R2)	0.9997	0.9996
Mean square error (MSE)	0.0012	0.0015
Mean absolute error (MAE)	0.098	0.0206

CONCLUSIONS

Owing to the gained results of this contribution the following major conclusions can be drawn.

- Adequate agreement between gain dew point pressure from the developed intelligent model and corresponding real recovery factor/net present value (NPV) values is observed. In other words, the conventional approaches fail to monitor real recovery factor/ net present value (NPV) of chemical flooding dedicated to the gained statistical criteria such as mean square error (MSE) and correlation coefficient.

- The evolved intelligent network model for monitoring real recovery factor/net present value (NPV) of chemical flooding is user friendly, fast, and cheap for implementation. Moreover, it is very useful and user friendly for evolving the accuracy and robustness of the commercial reservoir simulators like ECLIPSE and computer modelling group (CMG) software for enhanced oil recovery (EOR) from oil reservoirs.

REFERENCES

1. B. Shaker Shiran and A. Skauge, "Enhanced oil recovery (EOR) by combined low salinity water/polymer flooding," Energy and Fuels, vol. 27, no. 3, pp. 1223–1235, 2013.

2. S. Strand, T. Puntervold, and T. Austad, "Effect of temperature on enhanced oil recovery from mixed-wet chalk cores by spontaneous imbibition and forced displacement using seawater," Energy and Fuels, vol. 22, no. 5, pp. 3222–3225, 2008.

3. H. Zhang, M. Dong, and S. Zhao, "Which one is more important in chemical flooding for enhanced court heavy oil recovery, lowering interfacial tension or reducing water mobility?" Energy & Fuels, vol. 24, no. 3, pp. 1829–1836, 2010.

4. A. Z. Abidin, T. Puspasari, and W. A. Nugroho, "Polymers for enhanced oil recovery technology,"Procedia Chemistry, vol. 4, pp. 11–16, 2012.

5. M. A. Ahmadi, Y. Arabsahebi, S. R. Shadizadeh, and S. Shokrollahzadeh Behbahani, "Preliminary evaluation of mulberry leaf-derived surfactant on interfacial tension in an oil-aqueous system: EOR application," Fuel, vol. 117, pp. 749–755, 2014.

6. P. Luo, Y. Zhang, and S. Huang, "A promising chemical-augmented WAG process for enhanced heavy oil recovery," Fuel, vol. 104, pp. 333–341, 2013.

7. A. Mandal, A. Samanta, A. Bera, and K. Ojha, "Characterization of oil-water emulsion and its use in enhanced oil recovery," Industrial and Engineering Chemistry Research, vol. 49, no. 24, pp. 12756–12761, 2010.

8. A. Sabhapondit, A. Borthakur, and I. Haque, "Water soluble acrylamidomethyl propane sulfonate (AMPS) copolymer as an enhanced oil recovery chemical," Energy and Fuels, vol. 17, no. 3, pp. 683–688, 2003.

9. M. S. Karambeigi, R. Zabihi, and Z. Hekmat, "Neuro-simulation modeling of chemical flooding,"Journal of Petroleum Science and Engineering, vol. 78, no. 2, pp. 208–219, 2011.

10. M. A. Ahmadi and S. R. Shadizadeh, "Implementation of a high-performance surfactant for enhanced oil recovery from carbonate reservoirs," Journal of Petroleum Science and Engineering, vol. 110, pp. 66–73, 2013.

11. G. Chen, X. Wang, Z. Liang, et al., "Tontiwachwuthikul P: simulation of CO_2-oil Minimum Miscibility Pressure (MMP) for CO_2 Enhanced Oil Recovery (EOR) using neural networks," Energy Procedia, vol. 37, pp. 6877–6884, 2013.

12. N. Loahardjo, X. Xie, and N. R. Morrow, "Oil recovery by sequential waterflooding of mixed-wet sandstone and limestone," Energy & Fuels, vol. 24, no. 9, pp. 5073–5080, 2010.

13. M. Ali Ahmadi and M. Golshadi, "Neural network based swarm concept for prediction asphaltene precipitation due to natural depletion," Journal of Petroleum Science and Engineering, vol. 98-99, pp. 40–49, 2012.

14. M. A. Ahmadi, "Neural network based unified particle swarm optimization for prediction of asphaltene precipitation," Fluid Phase Equilibria, vol. 314, pp. 46–51, 2012.

15. S. Zendehboudi, M. A. Ahmadi, L. James, and I. Chatzis, "Prediction of condensate-to-gas ratio for retrograde gas condensate reservoirs using artificial neural network with particle swarm optimization," Energy & Fuels, vol. 26, no. 6, pp. 3432–3447, 2012.

16. M. Ali Ahmadi, S. Zendehboudi, A. Lohi, A. Elkamel, and I. Chatzis, "Reservoir permeability prediction by neural networks combined with hybrid genetic algorithm and particle swarm optimization," Geophysical Prospecting, vol. 61, no. 3, pp. 582–598, 2013.

17. S. Zendehboudi, M. A. Ahmadi, O. Mohammadzadeh, A. Bahadori, and I. Chatzis, "Thermodynamic investigation of asphaltene precipitation during primary oil production: laboratory and smart technique," Industrial and Engineering Chemistry Research, vol. 52, no. 17, pp. 6009–6031, 2013.

18. S. Zendehboudi, M. A. Ahmadi, A. Bahadori, A. Shafiei, and T. Babadagli, "A developed smart technique to predict minimum miscible pressure—EOR implications," Canadian Journal of Chemical Engineering, vol. 91, no. 7, pp. 1325–1337, 2013.

19. M. A. Ahmadi and S. R. Shadizadeh, "New approach for prediction of asphaltene precipitation due to natural depletion by using evolutionary algorithm concept," Fuel, vol. 102, pp. 716–723, 2012.

20. M. A. Ahmadi, "Prediction of asphaltene precipitation using artificial neural network optimized by imperialist competitive algorithm," Journal of Petroleum Exploration and Production Technology, vol. 1, no. 2–4, pp. 99–106, 2011.

21. S. Zendehboudi, A. R. Rajabzadeh, A. Bahadori et al., "Connectionist model to estimate performance of steam-assisted gravity drainage in fractured and unfractured petroleum reservoirs: enhanced oil recovery implications," Industrial and Engineering Chemistry Research, vol. 53, no. 4, pp. 1645–1662, 2014.

22. M. A. Ahmadi, M. Ebadi, and S. M. Hosseini, "Prediction breakthrough time of water coning in the fractured reservoirs by implementing low parameter support vector machine approach," Fuel, vol. 117, part A, pp. 579–589, 2014.

23. M. A. Ahmadi, M. Ebadi, A. Shokrollahi, and S. M. J. Majidi, "Evolving artificial neural network and imperialist competitive algorithm for prediction oil flow rate of the reservoir," Applied Soft Computing Journal, vol. 13, no. 2, pp. 1085–1098, 2013

24. M. A. Ahmadi and M. Ebadi, "Evolving smart approach for determination dew point pressure through condensate gas reservoirs," Fuel, vol. 117, part B, pp. 1074–1084, 2014.

25. M. A. Ahmadi, R. Soleimani, and A. Bahadori, "A computational intelligence scheme for prediction equilibrium water dew point of natural gas in TEG dehydration systems," Fuel, vol. 137, pp. 145–154, 2014.

26. M. A. Ahmadi, M. Ebadi, and A. Yazdanpanah, "Robust intelligent tool for estimating dew point pressure in retrograded condensate gas reservoirs: application of particle swarm optimization,"Journal of Petroleum Science and Engineering, 2014.

27. J. Kennedy, "Particle swarm: social adaptation of knowledge," in Proceedings of the IEEE International Conference on Evolutionary Computation (ICEC '97), pp. 303–308, Indianapolis, Ind, USA, April 1997.

28. M.-A. Ahmadi, M. Masumi, R. Kharrat, and A. H. Mohammadi, "Gas analysis by in situ combustion in heavy-oil recovery process: experimental and modeling studies," Chemical Engineering & Technology, vol. 37, no. 3, pp. 409–418, 2014.

29. G. V. Cybenko, "Approximation by superpositions of a sigmoidal function," Mathematics of Control, Signals, and Systems, vol. 2, no. 4, pp. 303–314, 1989.

Chapter 2

Life Cycle Assessment and Life Cycle Cost of Waste Management—Plastic Cable Waste

Mats Zackrisson, Christina Jönsson, Elisabeth Olsson

Energy and Environment Group, Department of Materials, Swerea IVF AB, Mölndal, and Sweden

ABSTRACT

The main driver for recycling cable wastes is the high value of the conducting metal, while the plastic with its lower value is often neglected. New improved cable plastic recycling routes can provide both economic and environmental incentive to cable producers for moving up the "cable plastic waste ladder". Cradle-to-gate life cycle assessment, LCA, of the waste management of the cable scrap is suggested and explained as a method to analyze the pros and cons of different cable scrap recycling options at hand. Economic and environmental data about different recycling processes and other relevant processes and materials are given. Cable producers can use this data and method to assess the way they deal with the cable plastic waste today and compare it with available alternatives and thus illuminate the improvement potential of recycling cable plastic waste both in an environmental and in an economic sense. The methodology applied consists of: cradle-to-gate LCA for waste material to a recycled material (recyclate); quantifying the climate impact for each step on the waste ladder for the specific waste material; the use of economic and climate impact data in parallel; climate impact presented as a span to portray the insecurities related to which material the waste will replace; and possibilities for do-it-yourself calculations. Potentially, the methodology can be useful also for other waste materials in the future.

INTRODUCTION

The main driver for recycling cables is the high value of the conducting metal (usually copper or aluminium), while the plastic with its lower value is often neglected. On the other hand, if it is not for the metal, the whole waste cable may be neglected, as it is often experienced with optical waste cables today [1]. This paper aims to provide primarily cable producers with a methodology to assess the way they deal with the cable plastic waste today and compare it with available alternatives and thus facilitate realizing the improvement potential of recycling cable plastic waste. The hypothesis is that it is possible to create a transparent methodology that provides additional insights and incentives of the value of recycling the plastic parts in addition to the metal core of the cable. Through using the methodology provided, the reader or user will be able to show the climate effects of improving the cable waste recycling (compared to how it is done today) and also to show the economic, technical and management implications of such improvements. The methodology as such can also be applicable to other waste materials.

The situation with small or negligible profit margins is similar for many other waste materials, for example textiles and construction waste [2]. It is therefore important to include economic data when analyzing waste recycling options. The simple knowledge that recycling a particular waste would lead to reduced environmental impact will not automatically lead to that it will be done; it will have to be economically beneficial (or enforced by law) otherwise it will not happen.

This paper and underlying report [3] have been compiled within the scope of the Wire and Cable project which is managed by the Swedish research institute Swerea IVF and financed by Vinnova, a Swedish governmental funding agency, and participating companies. The following cable manufacturers, polymer manufacturers, cable users and recycling companies are members of the project running from 2010 to 2013: Borealis AB, Draka Kabel Sverige AB, Ineos ChlorVinyls, Nexans Sweden AB, ABB AB, Stena Metall AB, Volvo Lastvagnar AB, Volvo Personvagnar AB and Ericsson AB. The main objective of the Cable project is to facilitate increased recycling of cable plastics. As from 2014, the Wire and Cable project will continue with many of the old members and some new ones.

METHOD

The methodology described and applied below consists of: cradle-to-gate LCA, life cycle assessment, for waste material to recyclate; quantifying the climate impact for each step on the waste ladder for the specific waste material; the reporting of economic and climate impact data in parallel; presentation of the climate impact as a span to portray the insecurities related to which material the waste will replace; and possibilities for do-it-yourself calculations. It has been developed in cooperation with the cable industry and used by them. It follows guidance about LCA of waste management issued by EUs Joint Research Centre [4] , which in turn builds on the International Organization for Standardization (ISO) 14044 standard [5] for LCA and the International Reference Life Cycle Data System Handbook [6] . Similar methodology may be useful also for waste categories other than cable waste.

Life Cycle Assessment in General

LCA according to ISO 14044 [5] consist of four stages: scooping, inventory, environmental impact assessment and interpretation. All stages except the one for environmental impact assessment are considered obligatory. The stages are often repeated in an iterative way that gradually refines the assessment. None of the stages are unique to the LCA methodology. What makes LCA unique is that all (or as many as possible/relevant) life cycle phases of the analyzed object are included from raw material extraction to the product's end-of-life [7] . The life cycle phases are often referred to as raw material production, (own) manufacturing, use and end-of-life [8] , see Figure 1.

When all life cycle phases are included in an LCA study, it is referred to as a cradle-to-grave study [4] . Studies that only include data about raw material production and own manufacturing are referred to as cradle-togate studies. Such cradle-to-gate LCA studies exist for most commodities like different steels, plastics etc.

Proposed LCA Application

Life cycle assessment of waste materials or waste management, though very common, has no special name in literature. A complete product life cycle as depicted in Figure 1 is rarely involved [4] . Instead, focus lies on recycling processes after the use phase or directly after the manufacturing processes as shown in Figure 2. Also production of virgin materials is included in order to account for that material recycling avoids primary or virgin material production. As can be seen, LCA of waste materials span over two adjacent product life cycles.

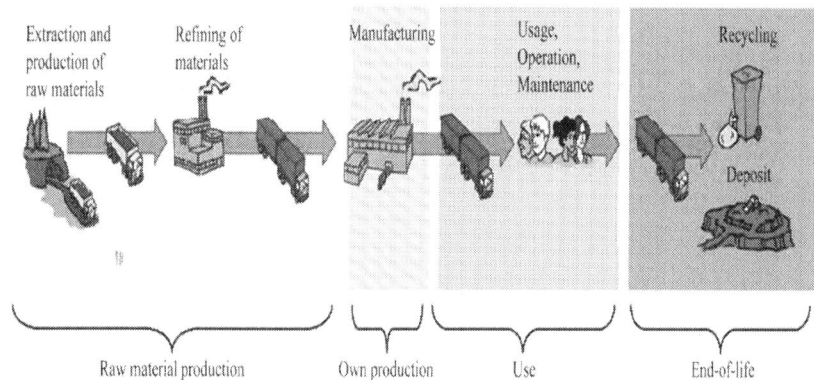

Figure 1. Life cycle assessment.

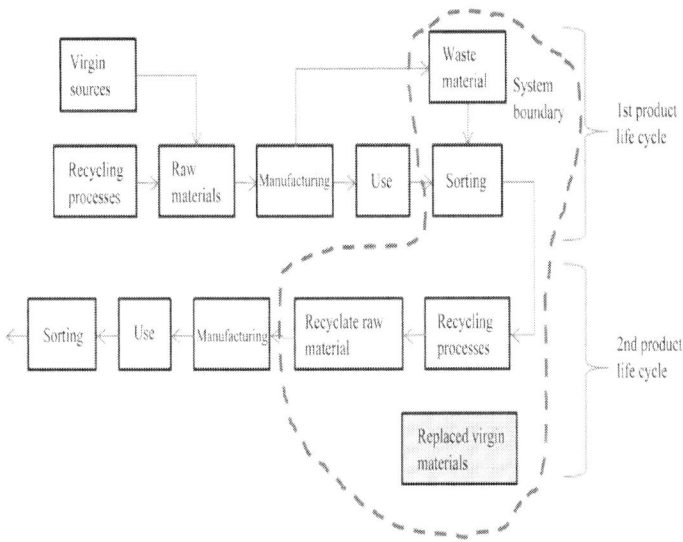

Figure 2. System boundary of LCA recycling study.

Often these product life cycles are for different products, i.e. cable plastic waste is rarely used to make new cables but rather to make other products. This is referred to as open-loop recycling, or, since it often entails a loss of valuable material properties, down-cycling [4]

Another way of seeing it is that the product or service under investigation is not the cable but rather the waste management of the cable, where the waste material is the input and the produced recyclate is the output, see Figure 3. In such a perspective the study could be compared to a cradle-to-gate LCA for a commodity from virgin origin, see Figure 1.

Focusing on the service required to manage the waste in the best way makes it natural to present the results per unit of cable plastic waste or per unit of cable waste, i.e. sometimes including the conducting metal. In LCA language these are the functional units used. A correct generic name of the functional unit would be waste management per unit of waste material. The starting point is the waste. Something has to be done about it; it cannot just be left in a pile; it has to be managed.

The waste ladder in Figure 4 portrays the waste management options generally available. It is considered in general to be environmentally preferable to be as high on the ladder as possible. The waste ladder or waste hierarchy is encouraged by the European Union (EU) Waste Framework Directive [9] , though departing from the hierarchy could be justified for, among other, reasons of technical feasibility and economic viability. In this paper, the climate impact associated with each step is calculated for the management of plastic waste from cables in order to further stimulate companies to move up the waste ladder.

The choice of system boundary and functional unit(s) means that there is no need to include the actual cable manufacturing or the use of the cables in the calculations. This is of course very advantageous since it limits drastically the amount of data needed for the analysis. However, such upstream activities may only be excluded if they are not affected by any of the investigated waste management options [4] . For example, if internal recycling of production cable waste is one option, the quality must be the same (as the normally used material) so that the cable manufacturing or use is not affected.

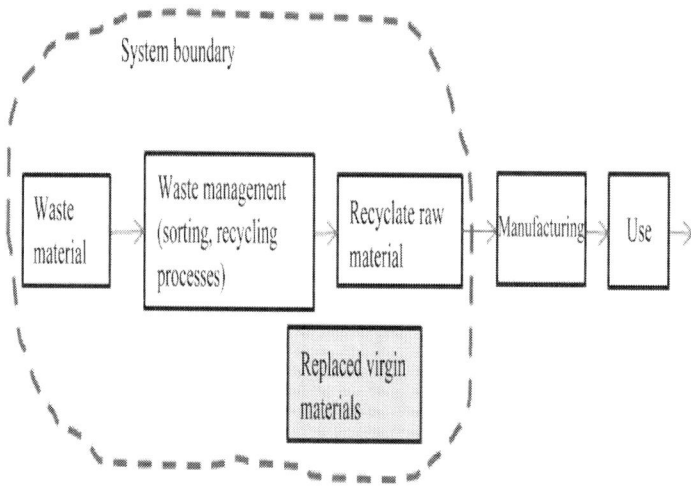

Figure 3: System boundary of cradle-to-gate waste management LCA.

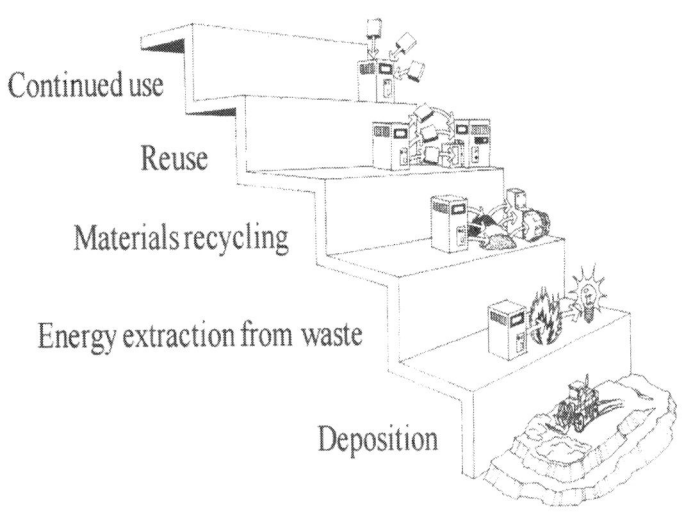

Continued use

Reuse

Materials recycling

Energy extraction from waste

Deposition

Figure 4: The waste ladder.

Cable producers and recycling companies have provided site-specific data for this paper and underlying reports [3] [10] . Sometimes averages from several companies with the same process are provided. For certain processes, e.g. transports and primary metal manufacturing, generic instead of specific process data is presented. This generic data stem from public LCA databases and represent in general European or global averages. Data has mainly been drawn from the database Ecoinvent 2.0 [11] .

The studied system is expanded to include avoided processes and the subsequent avoided environmental impacts when the recycled materials replace virgin materials. Avoided processes are shaded grey in the figures above. The choice of avoided process is very critical when system expansion is used [4] [7] . For example, since recycled copper can replace virgin copper, the environmental burdens for virgin copper manufacturing are subtracted from the studied system. Similarly, plastic recyclate can replace virgin plastic in some applications. However, it is rare that plastic recyclate can actually replace virgin plastic fully, more often some form of loss of material properties occur and this makes it difficult to define the replaced material. ISO 14044 [5] contains guidance concerning deciding the avoided or replaced material. The replacement should be based on, in priority order:

- Physical properties (e.g. mass or energy content);
- Economic value (e.g. market value of the scrap material or recycled material in relation to market value of primary material).

In order to follow ISO 14044 and also to somewhat capture that recycling almost always entail down-cycling, i.e., the recyclate has less good properties than the virgin resource, this paper presents two different scenarios:

- 1 to 1. In the 1 to 1 scenario 1 kg of recyclate is replacing 1 kg of the virgin material; e.g. 1 kg of recyclate polyvinyl chloride (PVC) compound replaces 1 kg of virgin PVC compound. Thus, the environmental burdens associated with manufacturing of virgin PVC compound is subtracted from the studied system (the cradle-to-gate waste management LCA) on a 1 to 1 basis.

- Market based. In the market based scenario, it is assumed that the loss of quality of the recyclate is proportional to the relation between the price paid for the recyclate and the price paid for the virgin material. Due to the loss of quality the recyclate cannot replace virgin material of the same type. What it can replace we do not know, so we assume, that the "environmental burdens saved" are proportional to the loss of quality which we assume is proportional to the difference in price. For example, if PVC recyclate is paid at 88 euro (EUR) per tonne and the price of virgin PVC is 1320 EUR per tonne, 88/1320 = 0.07 = 7% of the environmental burdens of virgin PVC manufacturing are subtracted from the studied system to account for the material that the recyclate is replacing.

Normally in LCA, at least five impact categories are used: climate change; acidification; ozone depletion; photochemical smog and eutrophication. In this paper, only climate change results are presented, in carbon dioxide equivalents (CO_2eq). Since CO_2eq is a good indicator for energy related environmental impact and most data sets related to cable waste management is dominated by energy use it has been shown [3] that CO_2eq is a good indicator for the environmental impact of cable waste recycling. For calculation of non CO_2 gases to CO_2eq the latest characterization factors from the Intergovernmental Panel on Climate Change (IPCC) [12] have been used.

Life Cycle Cost Methodology

Life cycle cost (LCC) [13] as applied in this paper refers to all costs and incomes of a particular waste incurred to the owner of the waste. For example, cost of transport (internal and external), waste processing costs, disposal costs and price of recyclates. The LCC thus shows (the waste owner) the profitability of each studied waste management option. The collection of data and identification of data required were achieved in parallel with the LCA.

RESULTS

In order to calculate the benefits of different recycling routes, monetary and climate change data about the involved recycling processes, transports and avoided (replaced) products have been collected. The units are euro (EUR) and gram carbon dioxide equivalents (CO_2eq). In order to facilitate calculations and comparisons, data have been assembled in one sheet, see Table 1. The idea with the sheet is to facilitate finding and marking data that is relevant for a unique comparison, sum it up and arrive at the results.

Reduction of Climate Impacts by Moving Up the Cable Plastic Waste Ladder

The potential gains of improving the cable plastic waste recycling by moving up the "waste ladder" are illustrated in Figure 5. All values are per kg of plastic waste. Note that the potential gains by the associated metal recycling are not shown in this figure.

The spans given reflect the different scenarios—market based or "1 to 1"—and the different polymers involved. Apart from the landfill figure, all figures are related to avoided products/processes, in the green part of Table 1. It is the replaced or avoided product that gives the largest climate impact contribution. Recycling processes and transports give only minor climate impacts and are therefore not included in the Figure 5 calculations. All figures are absolute, i.e. relative to zero.

Avoiding plastic (and metal) cable waste completely is of course the primary target of all cable producers, but not always possible. On the other hand, moving one or two steps up the ladder is not only possible but also to a degree driven by legislation limiting landfill and energy recycling. Moving one step up the waste ladder would mean avoiding around 0.5 kg CO_2eq/kg plastic. Consumption of plastic compounds by the European cable industry in 2012 was 1.23 million tonnes [14]. Plastic waste from cable manufacturers are around 5% of their total use of plastic [3]. If the industry as a whole can move one step up the waste ladder about $1,230,000,000 \times 0.05 \times 0.5$ kg CO_2eq/kg = 30,750 tonnes of CO_2eq can be avoided annually.

Life Cycle Cost or Economic Feasibility

Recycling processes and transports may only have minor climate impacts, but they are very important from an economical point of view. Therefore indicative price information is given in Table 1, in order to investigate the economic feasibility of different recycling options. Results from the LCC show that moving one step up at the top of the waste ladder can increase profits by almost 2 EUR per kg plastic, see Table 1. At the bottom of the plastic waste ladder there is a landfill cost of 0.12 EUR/kg and additional transportation costs.

Table 1. Hot milling of PVC scrap compared to external recycling

New recycling route	Hot milling		Compared to	Current way of recycling	External recycling		
Process(es)/ materials	Per kg cable		Comment/calculation	Process(es)/ materials	Per kg cable		Comment/calculation
	€	gram CO₂eq			€	gram CO₂eq	
Needed processes				Needed processes			
Production waste cable granulation	0.14	26		Production waste cable granulation	0.14	26	
Needed processes	Per kg plastic		Note that values below are given per kg polymer and may need recalculation to per kg cable by multiplication with value for kg polymer per kg cable.	Needed processes	Per kg plastic		Note that values below are given per kg polymer and may need recalculation to per kg cable by multiplication with value for kg polymer per kg cable.
	€	gram CO₂eq			€	gram CO₂eq	
Plastsep, Swedish electricity	0.055	2.2		Plastsep, Swedish electricity	0.055	2.2	
Plastsep, European electricity	0.055	8.9		Plastsep, European electricity	0.055	8.9	
Compounding PVC	0.23	6.2		Compounding PVC	0.23	6.2	
Compounding polyolefin	0.23	15		Compounding polyolefin	0.23	15	
Melt filtrating PVC	0.33	6.2		Melt filtrating PVC	0.33	6.2	
Melt filtrating polyolefin	0.33	15		Melt filtrating polyolefin	0.33	15	
Hot milling	0.004	4.3		Hot milling	0.004	4.3	
Disposal, polyvinylchloride	0.12	66		Disposal, polyvinylchloride	0.12	66	
Disposal, plastics, mixture	0.12	90		Disposal, plastics, mixture	0.12	90	
Needed transports	Per tonkm transport		Multiply values below for € and gram CO₂eq per tonkm transport with actual transport distance to get gram CO₂eq/tonne and €/tonne for the actual transport. Convert to gram CO₂eq /kg and €/kg by dividing with 1000.	Needed transports	Per tonkm transport		Multiply values below for € and gram CO₂eq per tonkm transport with actual transport distance to get gram CO₂eq/tonne and €/tonne for the actual transport. Convert to gram CO₂eq/kg and €/kg by dividing with 1000.
	€	gram CO₂eq			€	gram CO₂eq	
Lorry, Trailer 26t Euro3, NTM	0.10	51		Lorry, Trailer 26 t Euro3, NTM	0.02	10	200 km transport to granulation and compounding: 200 × 0.1/1000 = 0.02 euro and 200 × 51/1000 = 10 gram CO₂eq
Transport, lorry >16t, fleet average/ RER S	0.10	134		Transport, lorry >16 t, fleet average/RER S	0.07	94	700 km transport to user in Europe: 700 × 0.1/1000 = 0.07 euro and 700 × 51/1000 = 94 gram CO₂eq
Total needed processes	0.004	4.3		Total needed processes	0.460	137	

Avoided processes	Per kg material			Note that values below are given per kg material and may need recalculation to per kg cable by multiplication with value for kg material per kg cable.	Avoided processes	Per kg material			Note that values below are given per kg material and may need recalculation to per kg cable by multiplication with value for kg material per kg cable.
	€	Market 1 to 1	gram CO_2eq			€	Market 1 to 1	gram CO_2eq	
Copper, primary, at refinery/GLO S	-5.5	-3160	-3160		Copper, primary, at refinery/GLO S	-5.5	-3160	-3160	
Copper granulate	-5.3	-3065	-3160		Copper granulate	-5.3	-3065	-3160	
Copper fluff	-4.6	-2686	-3160		Copper fluff	-4.6	-2686	-3160	
Aluminium, primary, at plant/RER S	-1.6	-12,200	-12,200		Aluminium, primary, at plant/RER S	-1.6	-12,200	-12,200	
Aluminium granulate	-1.5	-11,346	-12,200		Aluminium granulate	-1.5	-11,346	-12,200	
Aluminium fluff	-1.3	-10370	-12,200		Aluminium fluff	-1.3	-10,370	-12,200	
Heavy fuel oil, at regional storage/RER S	-1.1	-455	-455		Heavy fuel oil, at regional storage/RER S	-1.1	-455	-455	
HFFR as oil replacement	0.058	9	-166		HFFR as oil replacement	0.058	9	-166	
Polyolefins as oil replacement	-0.022	-9	-477		Polyolefins as oil replacement	-0.022	-9	-477	
PVC compound for cable 1	-1.3	-1500	-1500		PVC compound for cable 1	-1.3	-1500	-1500	
PVC recyclate 1	-0.09	-100	-1500		PVC recyclate 1	-0.09	-100	-1500	
HFFR compound for cable	-2.0	-1170	-1170		HFFR compound for cable	-2.0	-1170	-1170	
HFFR recyclate	-0.11	-65	-1170		HFFR recyclate	-0.11	-65	-1170	
Compounding PVC	-0.23	-6.2	-6.2		Compounding PVC	-0.23	-6.2	-6.2	
Compounding polyolefin	-0.23	-15	-15		Compounding polyolefin	-0.23	-15	-15	

Avoided transports	Per ton km transport		Multiply values below for € and gram CO_2eq per tonkm transport with actual transport distance to get gram CO_2eq/tonne and €/tonne for the actual transport. Convert to gram CO_2eq /kg and €/kg by dividing with 1000.	Avoided transports	Per tonkm transport		Multiply values below for € and gram CO_2eq per tonkm transport with actual transport distance to get gram CO_2eq/tonne and €/tonne for the actual transport. Convert to gram CO_2eq /kg and €/kg by dividing with 1000.
	€	gram CO_2eq			€	gram CO_2eq	
Lorry, Trailer 26t Euro3, NTM	-0.02	-10	-10	200 km transport from PVC supplier avoided.	Lorry, Trailer 26 t Euro3, NTM	-0.10	-51	-51
Transport, lorry >16t, fleet average/RER S	-0.10	-134	-134		Transport, lorry > 16 t, fleet average/RER S	-0.10	-134	-134

	Per kg material					Per kg material		
	€	Market 1 to 1	gram CO_2eq			€	Market 1 to 1	gram CO_2eq
Total avoided processes	-1.550	-1516	-1516		Total avoided processes	-0.09	-100	-1500
Total of needed and avoided processes	-1.546	-1512	-1512		Total of needed and avoided processes	0.370	37	-1364

Bottom line comparison of new recycling route compared to current way of recycling

Processes / materials	€	Per kg Market 1 to 1 gram CO₂eq			Conclusions
New recycling route	-1.546	-1512	-1512	For 100 tonnes the savings are 100 × 1.92 × 1000 = 192 000 euro and in between 154.8 - 14.8 tonnes CO₂eq.
Current way of recycling	0.370	37	-1364	
Difference	-1.92	-1548	-148	

Notes to the calculation of the LCA comparison between Hot milling and External recycling in Figure 6 and Table 1. Boxes coloured yellow in table 1 apply! Hot milling, left columns: Hot milling costs 0.004 EUR/kg plastic and entails 4.3 gram CO_2eq/kg plastic emissions. Production of virgin PVC is avoided, thus 1.3 EUR/kg plastic and 1500 gram CO_2eq/kg plastic is avoided. Compounding of PVC is avoided, thus 0.23 EUR/kg plastic and 6.2 gram CO_2eq/kg plastic is avoided. Transport, 200 km, of virgin PVC is avoided, thus 0.1 × 200/1000 = 0.02 EUR/kg plastic and 51×200/1000=10 gram CO_2eq/ kg plastic is avoided. Since there is no down-cycling of the material (no loss of quality), the market perspective and the 1 to 1 perspective yield the same results! For Hot milling, the total of needed processes minus avoided processes is a gain of 1.546 EUR/kg plastic and avoidance of 1512 gram CO_2eq/kg plastic. External recycling, right columns: Granulation of hardened lumps costs 0.14 EUR/kg plastic and entails 26 gram CO_2eq/kg plastic emissions. Compounding the granulated PVC costs 0.23 EUR/kg plastic and entails 6.2 gram CO_2eq/kg plastic emissions. Transport, 200 km, of PVC lumps to granulation costs 0.1 × 200/1000 = 0.02 EUR/kg plastic and entails 51 × 200/1000 = 10 gram CO_2eq/kg plastic emissions. Transport, 700 km, of compunded PVC recyclate to user in Europe costs 0.1 × 700/1000 = 0.07 EUR/kg plastic and entails 134 × 700/1000 = 94 gram CO_2eq/kg plastic emissions. Reuse of PVC recyclate bring an income of 0.09 EUR/kg plastic and avoids 88/1320 × 1500 = 100 gram CO_2eq/kg plastic emissions in a market perspective and 1500 CO_2eq/kg plastic in a 1 to 1 perspective. For external recycling, the total of needed processes minus avoided processes is a loss of 0.37 EUR/kg plastic. The climate impact range between emissions of 37 gram CO_2eq/kg plastic and avoidance of 1364 gram CO_2eq/kg plastic. Bottom line: Employing hot milling instead of external recycling saves −1.546 − 0.370 = −1.92 EUR/kg plastic (minus sign means savings/income/avoidance) and avoids emissions between

−1511 − 37 = −1548 gram CO_2eq/kg plastic and −1512 − (−1364) = −148 gram CO_2eq/kg plastic.

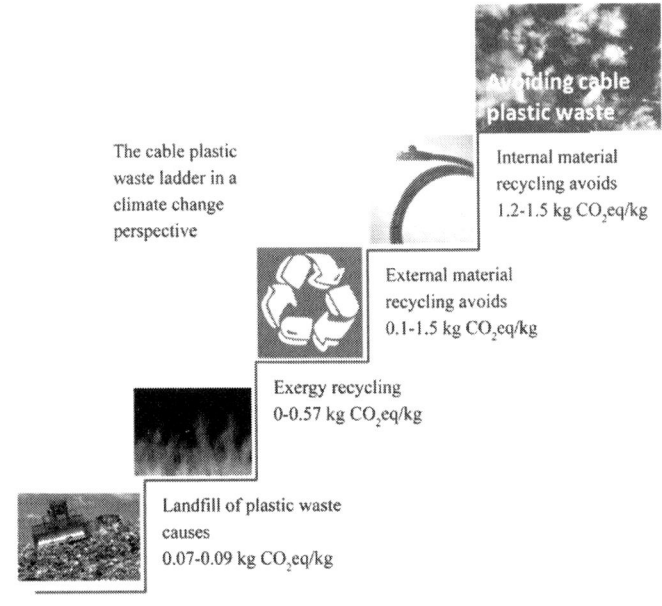

The cable plastic waste ladder in a climate change perspective

Avoiding cable plastic waste

Internal material recycling avoids 1.2-1.5 kg CO_2eq/kg

External material recycling avoids 0.1-1.5 kg CO_2eq/kg

Exergy recycling 0-0.57 kg CO_2eq/kg

Landfill of plastic waste causes 0.07-0.09 kg CO_2eq/kg

Figure 5. Potential climate change gains (per kg of plastic waste) by moving up the cable plastic waste ladder.

Calculation Example

Below is given an example of how to use the data in Table 1 to calculate the economics and the climate impact of a particular recycling case. It is always easier to understand the results if they are compared to one another. As a standard comparison, the current way of managing the cable plastic waste is used. The yellow boxes in Table 1 are summed up in the Total boxes. A clean sheet to perform other calculations is available in the underlying report [3] that also contains other calculation examples and a step-by-step procedure on how to perform them.

Internal versus External Recycling

Start and stop PVC and HFFR (halogen free flame retardant) scrap at cable extruders can be recycled directly back into extruders via hot milling of the scrap. This might need investment in mills which is not considered in this calculation. The extra work involved could often be handled by the extruder operator, thus, normally hot milling at extruders does not entail any extra work costs. It is difficult to clean the mills. Therefore, hot milling is only relevant at extruders that run the same material all the time. For increased meaning and understanding, hot milling at extruders should be compared to an alternative. A currently used alternative is to sell the hardened scrap lumps from the extruder to an external waste handler who granulates them and pass them on for mechanical recycling in a different product. It is a good idea to make a rough drawing of the processes involved in both recycling routes, see below. Avoided processes and materials are coulored yellow in Figure 6. Note that upstream processes like transport and compounding of virgin PVC is avoided by hot milling the scrap at extruders. A minus sign in Table 1 means saved or avoided EUR or CO_2eq. It is recommended to do the calculations both with a market perspective on the avoided burdens and with a "1 to 1" perspective.

When the table has been completed a comparison of the bottom lines for hot milling and external recycling is done. The conclusion, in the example, is that hot milling can save more than 192000 EUR annually and avoid between 15 - 155 tonnes of CO_2eq annually. Per kg, the figures compare well with those given for internal and external material recycling in the cable plastic waste ladder in Figure 5.

DISCUSSION

The methodology described and applied has evolved during a four year long cooperation between industrial waste management experts, the cable industry, LCA practitioners and other stakeholders. It aims to bring to the industrial decision-maker the necessary economic and environmental facts to judge the merits of competing waste management options and thus facilitate movement up the waste ladder. Some barriers and incentives related to this aim of improving waste management of plastic cable waste in particular and other waste are discussed below.

Restricted Substances May Hinder Recycling

It should be pointed out that the use in cables of restricted phthalates as well as restricted substances such as lead, bromide and antimony may hinder the possibilities to use recycled plastic waste from used cable. The inclusion of restricted substances in cables may also hinder all forms of external recycling of plastic waste from cable production. The subject of restricted substances in cable waste will be the focus in the Wire and Cable project run by Swerea IVF, from 2014 and onwards. The quality of various waste materials and the phasing out of hazardous waste is regulated under the Waste Framework Directive [9] , where End-of-Waste criteria is developed for priority waste streams, among them plastic materials. Chemical regulations like REACH (Registration, Evaluation, Authorisation and Restriction of Chemicals) [15] and product directives like RoHS (Restriction of Hazardous Substances) [16] regulates hazardous substances in products and material used in products, regardless virgin or recycled material. This potential barrier for recycling exists for all waste materials.

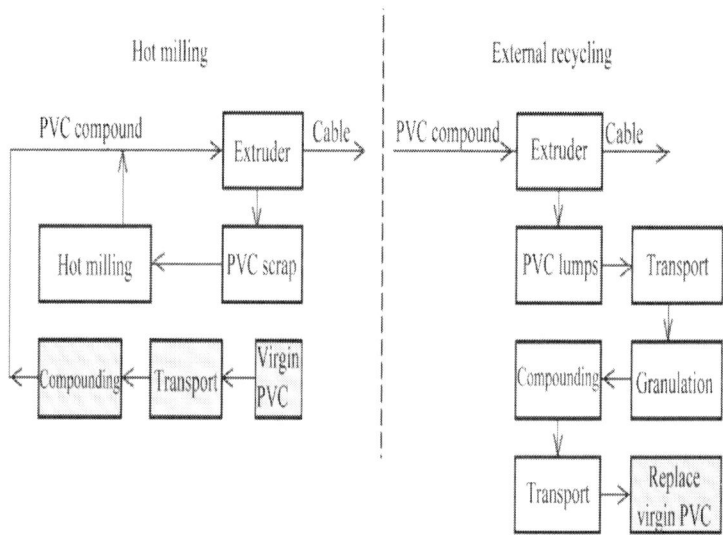

Figure 6. Hot milling of PVC scrap compared to external recycling of hardened PVC lumps.

Identical Processes

When comparing recycling routes that contain identical processes it is not necessary to include the identical processes since they equal out each other. However, when you carry out the analysis in practice it is easier to just list all the processes involved without thinking about if they equal out or not. Once included in the analysis, the processes carry some information, so you might as well leave them in. The recommendation is therefore to include also identical processes in a comparison once you have identified them. "Identical processes" that are initially forgotten need not be taken in to the calculations.

Limitations of the System Boundary

To be able to limit the analysis to include only the waste management processes is very advantageous since it limits drastically the amount of data needed for the analysis. One reason why this is often possible for cable waste is that when producing cables, achieving the specified product quality properties is far more important than advancing on the waste ladder. Thus, in practice, it can be assumed that the manufacturing and use of the cable will in general not be affected by any of the waste management options investigated; simply because the cable properties must remain the same. So such upstream activities can most often be excluded from the LCA. Seen from another perspective, achieving the specified product properties with the use of waste materials is probably the largest barrier to (internal) material recycling.

Downstream, the recommended use of two waste perspectives (market and 1 to 1), can be seen as a LCA sensitivity analysis regarding the market for the recyclate [4] . In general, the 1 to 1 approach overestimates the CO_2eq savings, while the market approach probably in most cases underestimates the CO_2eq savings. The true climate change influence most probably is somewhere in between these two extremes. It is therefore recommended to do both calculations and present the results as a span, as is done in the example calculations.

From an economic point of view the market approach must be used in order to understand the economic implications. PVC recyclate rarely sell at the same price as virgin PVC. However, when there is no price difference between the recyclate and the virgin material, then the market approach gives the same result as the 1 to 1 approach. Put another way, the market approach is only relevant when so called down-cycling of wastes occur, i.e., the waste is not reused for the same product/purpose as it was used originally or maintaining its origin quality.

Ecodesign of Cables

Within the Wire and Cable project, some attempts were made to elaborate guidelines for how to design cables in order to facilitate material recycling after their use. These attempts were however unsuccessful. One reason may be that in life cycle assessment of cables, as apart from LCA of cable recycling, the materials in the cable generally carry insignificant environmental impact compared to the environmental impact associated by the electricity losses during a cables life cycle [17] . In other words, the dominant environmental impact occurs during the use phase of a cable and therefore improvement work should focus on the use phase. Subsequently, less attention is put on the recycling stage. However, from a cable manufacturing point of view, the material costs are significant, and this should warrant some interest at least for improving the recyclability of production cable waste.

The Relevance of Climate Change

Plastsep is a technology based on sink-float separation and wet shaking table that is used to separate heavy plastics like PVC from the lighter polyolefines. Applying Plastsep to the mixed plastic waste output after granulation of used cables has proven both economically and environmentally beneficial under all conceivable circumstances [12] . It was further shown [12] that applying Plastsep not only avoided climate impact but also photochemical smog formation, eutrophication and acidification. This indicates that climate change can serve as an indicator also for these other environmental impact categories.

Prices

All prices should be seen as indicative. A variation of at least +- 100% should be taken for granted. Nevertheless, having access to price information of some kind is very beneficial since it is never a potential climate impact avoidance alone that will make companies move up the waste ladder. Moving up the waste ladder will have to be economically beneficial (or enforced by law) otherwise it will not happen. With the information given companies can get a first idea which recycling routes are economically viable for them and which are not.

Applicability of Methodology for Other Waste Materials

The applicability of the applied methodology to other waste materials than plastic cable waste is discussed below for each of the five steps in the methodology.

Cradle-to-Gate LCA (and LCC) For Waste Material to Recyclate

Data relevant to the particular waste material is needed on:
- Recycling processes in the form of energy use, emissions, yield, price etc.
- Transports in the form of energy use, emissions and price.
- Primary material production in the form of energy use, emissions, price.
- Price of recyclates.

Quantifying the Climate Impact for Each Step on the Waste Ladder

To accomplish a quantified waste ladder, the particular waste material need to be well defined and the market for the waste material known, i.e. prices of recyclates. The quantified waste ladder is a pedagogic instrument to be used to rouse interest. The specific case should always be calculated with specific data.

Use of Economic and Environmental Data in Parallel

Cost data is often very sensitive. The related price information could be much easier to get access to and it is also more relevant than the cost. Fluctuation of market prices has to be considered in order to achieve long-term sustainability of waste management options.

Climate Impact Presented As a Span

It may be that some other environmental impact category than climate impact is more relevant to use or that several impact categories are needed. This should be checked for all waste materials.

As discussed above, to calculate and present the climate impact as a span, can be seen as an LCA sensitivity analysis regarding the market for the recyclate. The true climate change influence most probably is somewhere in between the top value and the lower value. If there is no price difference between the recyclate and the virgin material, then the top or high value of CO_2eq savings apply.

Do-It-Yourself Calculations

The vision of laymen doing correct life cycle calculations may be difficult to realize, but aiming there will aid in understanding and communicating the results of the life cycle calculations. This will, for example, facilitate exchange and correct understanding of LCA data between LCA practitioners. The calculations (for cable waste) can be achieved by the following steps:

- Copy the table with all the process data and enlarge it to A3-size or larger.
- Identify the recycling routes you want to compare. Draw simple process flow diagrams of both recycling routes, from waste to recyclate.
- In the A3-sheet, identify and mark the (grey) processes needed to enable the recycling
- In the A3-sheet, identify and mark the (green) products/materials and processes that are avoided/replaced by the recyclate

- Convert data to per kg cable if necessary, see examples
- Convert transport data according to instructions in the table
- Examine the bottom lines of the recycling routes and calculate the difference in savings per kg 8) Calculate the difference in savings per year for your company

CONCLUSIONS

This paper suggests cradle-to-gate life cycle assessment as a method to analyze the pros and cons of different cable scrap recycling options at hand. Economic and environmental data about different recycling processes and other relevant processes and materials have been collected and are presented. Cable producers could use this data and the proposed method to assess the way they deal with the cable plastic waste today and compare it with available alternatives and thus illuminate the improvement potential of recycling cable plastic waste both in an environmental and in an economic sense. The methodology suggested and applied consists of:

- Cradle-to-gate LCA for waste material to recyclate;
- Quantifying the climate impact for each step on the waste ladder;
- Use of economic and climate impact data in parallel;
- Climate impact presented as a span; and 5) Do-it-yourself calculations.

In this paper, comparisons between internal and external recycling of cable plastics show that between 1.5 - 0.15 kg CO_2eq can be saved per kg of plastics when moving one step up the waste ladder from external recycling to internal recycling. In economic terms, this one step up at the top of the waste ladder can increase profits by almost 2 EUR per kg plastic.

The suggested method is probably applicable also for other waste materials in society in order to move towards improved use of finite resources.

REFERENCES

1.   Unger, N. and Oliver, G. (2008) Life Cycle Considerations about Optic Fibre Cable and Copper Cable Systems: A Case Study. Journal of Cleaner Production, 16, 1517-1525.http://dx.doi.org/10.1016/j.jclepro.2007.08.016

2.   JRCb (2011) Supporting Environmentally Sound Decisions for Construction and Demolition (C & D) Waste Management.

3.   Zackrisson, M. (2013). Recycling Production Cable Waste—Environmental and Economic Aspects. Swerea IVF Report 13003, Mölndal.

4.   JRCa (2011) Supporting Environmentally Sound Decisions for Waste Management. JRC European Commission.

5.   ISO (2006) ISO 14044. Environmental Management—Life Cycle Assessment—Requirements and Guidelines.

6.   Wolf, M.-A. and Rana, P. (2012) The International Reference Life Cycle Data System.http://eplca.jrc.ec.europa.eu/uploads/2014/02/JRC-Reference-Report-ILCD-Handbook-Towards-more-sustainable-production-and-consumption-for-a-resource-efficient-Europe.pdf

7.   Björklund, A. and Finnveden, G. (2005) Recycling Revisited-Life Cycle Comparisons of Global Warming Impact and Total Energy Use of Waste Management Strategies. Resources, Conservation and Recycling, 44, 309-317.http://dx.doi.org/10.1016/j.resconrec.2004.12.002

8.   Zackrisson, M., Cristina, R., Kim, C. and Anna, J. (2008) Stepwise Environmental Product Declarations: Ten SME Case Studies. Journal of Cleaner Production, 16, 1872-1886.http://dx.doi.org/10.1016/j.jclepro.2008.01.001

9.   European Commission (2008) Directive 2008/98/EC of the European Parliament and of the Council of 19 November 2008 on Waste and Repealing Certain Directives.

10.   Zackrisson, M. (2012) Life Cycle Assessment of Cable Recycling. Part I: Plastsep Compared to State of the Art. Swerea IVF Report, Swerea IVF AB, Mölndal.

11.   Ecoinvent (2010) The Life Cycle Inventory Data Version 2.2. Ecoinvent Database.

12. IPCC (2007) Climate Change 2007 the Physical Science Basis. http://www.ipcc.ch/publications_and_data/ar4/wg1/en/ch2s2-10-2.html#table-2-14

13. IEC (2004) Dependability Management—Part 3-3: Application Guide—Life Cycle Costing. IEC 60300-3-3.

14. AMI (2012) AMI's Guide to the Cable Extrusion Industry in Europe (Edition 6).www.amiplastics.com

15. European Commission (2006) REGULATION (EC) No 1907/2006 of the European Parliament and of the Council of 18 December 2006 Concerning the Registration, Evaluation, Authorisation and Restriction of Chemicals (REACH).

16. European Commission (2011) DIRECTIVE 2011/65/EU of the European Parliament and of the Council of 8 June 2011 on the Restriction of the Use of Certain Hazardous Substances in Electrical and Electronic Equipment.

17. Jones, C.I. and Marcelle, C.M. (2010) Life-Cycle Assessment of 11 kV Electrical Overhead Lines and Underground Cables. Journal of Cleaner Production, 18, 1464-1477.http://dx.doi.org/10.1016/j.jclepro.2010.05.008

Unsteady Fluidynamic Behavior of Gas Bubbles Flowing in Curved Pipes: A Numerical Study

Jose Luis Gomes Marinho[1], Ramdayal Swarnakar[1],
Severino Rodrigues de Farias Neto[1]
and Antonio Gilson Barbosa de Lima[2]

[1]Department of Chemical Engineering, Center of Sciences and Technology, Federal University of Campina Grande, Campina Grande, Brazil

[2]Department of Mechanical Engineering, Center of Sciences and Technology, Federal University of Campina Grande, Campina Grande, Brazil

ABSTRACT

Petroleum is considered as one of the factors for the development of a nation as well as a cause of economic and political conflicts around the world because of the diversity of products obtained with their derivatives such as fuel for automotives and aviation, and manufacturing plastic parts, among others. The crude petroleum (usually oil, water and gas) found in an underground reservoir is transported to the surface by pipes, and has drawn the attention of researchers because of the problems generated in the pipeline with particular attention to the loss of pressure, friction and bubbles. For a fluid flow in plug regime, where many of the bubbles formed coalesce and produce bigger ones of sizes almost equal to the pipe diameter (Taylor bubble), severe instability in the flow is caused. In this context, the objective of this research has been to study the Taylor bubble flow in curved ducts using the software CFX. Results of the transient effects of the air concentration on the bubble air volumetric fraction, of the viscosity on bubble format, and pipe angle of 90° on bubble symmetry are presented and interpreted.

INTRODUCTION

Petroleum is considered as an important factor for the development of a nation. It is extracted out of the underground reservoirs. It is generally composed of oil, air and water, which are carried up to the surface by pipes. To foresee the possible fluid flow instabilities in the pipes, researchers have been studying the behavior of the multiphase flow in the interior of pipes.

Slug flow is one of the most common and complex flow patterns in two-phase flow characterized by long gas bubbles almost filling the pipe cross-section, where liquid moves around the bubbles and in bulk between two successive bubbles. Slug flow exists over a broad range of gas and liquid flow rates and is encountered in a wide variety of industrial applications like oil and gas wells, process vaporizers and gas-liquid pipeline reactors [1-6].

The study of bubble behaviors, mainly that which consists of elongated bubbles, is one of the important aspects associated with two-phase flows.

Some authors who have reported, in literature, their studies on Taylor bubble behavior are [1-4] and [7-14].

Taylor bubble formation as the most dominant twophase flow pattern, in the miniature channels with stagnant liquids, has been reported by [13].

[14] proposed that elongated Taylor bubble can be divided into three parts according to the profile configuration of the acting forces (inertial forces, surface tension and viscosity). The first part is prolate

spherical cap zone, the third part is the terminal cylinder zone with terminal constant thickness and velocity of fully developed falling liquid film. The second part is the transition zone between the prolate spherical cap and the terminal cylinder. The results of viscosity effect are interesting, with a significant effect on the streamlines in the Taylor bubble wake zone. It was found that the higher the viscosity, the lesser is the distortion and the smaller is the fluctuation of the bubble bottom. These authors have also observed the presence of small bubbles in the Taylor bubble tail (oblate spheroidal part).

The classical Taylor bubble often observed in the laboratory usually results from air rising in water and has a prolate spheroidal leading edge and a flat, or even concave, trailing edge, [1]. According to these authors, the Taylor bubble formation occurs when the gas flow increases in the system, increasing the number of bubbles that form and there is a tendency to coalesce and to form bubbles with dimensions close to the duct diameter.

[10] found that the pipe wall did not influence the shape or diameter of the breaking bubble. This is because the bubble breaking was influenced by the buoyancy force and surface tension between bubble and the nozzle. However, once the bubble got disconnected from the nozzle, the pipe wall influenced the bubble behavior, being reflected in the slow rise of the bubble in the tube. The experimental results for the larger diameter tubes (D = 6.35 and 4.36 mm) indicated that the departure diameter of bubble is not affected by the wall of the tube, but the velocities of the bubble rising in large diameter tube were higher than those obtained with the tube of D = 3.18 mm.

An experimental study for different stages of bubble formation, through a point of air injection, has been reported by [10]. The injection nozzle had an internal diameter of 0.556 mm. The pipe diameter ranging from 1.89 to 6.35 mm and length from 200 mm to 270 mm were used. These authors studied bubble formation with the aid of a high-resolution camera, in circular, triangular, quadratic and rectangular pipes. It was observed that as more gas is injected bubble tends to grow more, and its initial spherical shape changed to an elliptical shape. This behavior is due to buoyancy forces. As the bubble grows it narrows at the injection nozzle and the contact from nozzle is broken taking the shape of a nearly perfect sphere in the vertical direction.

[2] published an experimental investigation of flow patterns and characteristics of two-phase flow in upward inclined tubes of 2 - 8 mm diameter. These authors concluded that the shape and the radial position of the gas slug, in the tube, are influenced by the tube diameter, flow rate and inclination angle of tube. Furthermore, the length of gas slug increases with the superficial gas velocity. The length of gas slug in the inclined tube is longer than that in the vertical or horizontal tube, and the gas slug velocity in the inclined tube is faster than that in the vertical or horizontal tube. Observations made under various inclinations of tubes showed that the flow pattern was elongated bubble flow and no small dispersed bubbles existed in liquid slug.

[12], in their review paper, have reported CFD studies of Taylor bubbles in 3D and 2D geometries. The objective was to study slug flow in micro-channels (0.25, 0.5, 0.75, 1, 2 and 3 mm). They mentioned that in the computational region, the slug length slightly increases with the increase of surface tension. However, there is almost no influence of liquid viscosity. It was observed that there is no significant difference in the slug length obtained when 3D and 2D geometries were studied. These authors concluded that: 1) gas slug length increases by increasing the superficial gas velocity, and by decreasing the superficial liquid velocity; 2) liquid slug length increases by increasing the superficial liquid velocity, and decreasing superficial gas velocity.

The study on the flow regime for two-phase gas-liquid flow in an inclined tube with small diameter is still very little in literature. Therefore, in order to make a contribution in this area of knowledge, present research aimed to study the unsteady behavior of gas bubbles flowing in curved pipes (90° angle), with particular reference to Taylor bubbles, using the software CFX-3D.

MATHEMATICAL MODELING

Governing Equations

To study the two-phase flow (gas-oil) in curved pipes (Figure 1), following conditions were adopted: 1) twodimensional field domain in cylindrical coordinates; 2) isothermal flow; 3) no chemical reaction; 4) phases treated as incompressible fluid with physical properties being constant; 5) no gravity effect; and 6) no interfacial mass transfer. Thus, the conservation equations are described as:

- Equation of mass conservation;

$$\frac{\partial}{\partial t}\left(f_\alpha \rho_\alpha\right)+\nabla\cdot\left(f_\alpha \rho_\alpha U_\alpha\right)=0$$

1)

- Equation of momentum conservation;

$$\frac{\partial}{\partial t}\left(f_\alpha \rho_\alpha U_\alpha\right)+\nabla\cdot\left[f_\alpha\left(\rho_\alpha U_\alpha \otimes U_\alpha\right)\right]$$

$$=-f_\alpha\nabla P+\nabla\cdot\left\{f_\alpha \mu_\alpha\left[\nabla U_\alpha+\left(\nabla U_\alpha\right)^T\right]\right\}$$

$$+f_\alpha g\left(\rho_\alpha-\rho_{ref}\right)+C_{\alpha\beta}\left(U_\beta-U_\alpha\right)+\rho g$$

(2)

where $C_{\alpha\beta}$ corresponds the interfacial drag term, given by:

$$C_{\alpha\beta}=\frac{C_D}{8}A_{\alpha\beta}\rho_\alpha\left|U_\beta-U_\alpha\right|$$

(3)

The indices α and β represent the continuous and dispersed phase; f, ρ, μ, U are the volume fraction, density, dynamic viscosity and velocity vector respectively, P is the pressure, and $A_{\alpha\beta}$ represents the density of interfacial area, which is given by:

$$A_{\alpha\beta}=\frac{6f_\beta}{d_\beta}$$

(4)

where f_β and d_β are volumetric fraction and air bubble diameter respectively.

The drag coefficient C_D was estimated by the Grace model [16]. This model considers the dispersed phase shape effect, a constant interfacial tension of the bubble and is given by:

$$C_D = \frac{4gd\,\Delta\rho}{3U_T^2\rho_c}$$

(5)

where U_T is terminal velocity of a rising bubble, $\Delta\rho$ is density difference between phases, d is bubble diameter and ρ_c is density of continuous phase.

In the model, a constraint equation was used, where the sum of the volumetric fraction of the phases is unity. This equation is given by:

$$\sum_{\beta=1}^{N_P} f_\alpha = 1$$

(6)

The pressure field used was same for all the phases. In this case it is given by:

$$P_\alpha = P_1 = P \text{ for } 2 \le \alpha \le N_p$$

(7)

Initial and Boundary Conditions

• Initial conditions:

At time t = 0, the pipe was completely full with oil, at pressure 101,325 Pa, and the velocity components of the two phases were considered void.

• Boundary conditions:

At inlet:

$$0 < r \leq R - \Delta r \begin{cases} U_z^{Gas} = U_{Max}\left(1 - \dfrac{r}{R}\right)^{1/10} \\ U_{Max} = 0.1 \text{ m/s} \\ U_z^{Liquid} = 0.0 \text{ m/s} \text{ for } t < t_{inj} \\ U_r^{Gas} = U_r^{Liquid} = 0.0 \text{ m/s} \\ f_{Gas} = 1.0 \text{ and } f_{Liquid} = 0.0 \end{cases}$$

$$R - \Delta r < r \leq R \begin{cases} U_z^{Gas} = 0 \text{ m/s} \\ U_z^{Liquid} = U_0 \\ U_r^{Gas} = U_r^{Liquid} = 0 \text{ m/s} \text{ for } t < t_{inj} \\ f_{Gas} = 0.0 \text{ and } f_{Liquid} = 1.0 \end{cases}$$

$$0 < r \leq R \begin{cases} U_z^{Gas} = 0.0 \text{ m/s} \\ U_z^{Liquid} = U_0 \text{ for } t > t_{inj} \\ U_r^{Gas} = U_r^{Liquid} = 0.0 \text{ m/s} \\ f_{Gas} = 0.0 \text{ and } f_{Liquid} = 1.0 \end{cases}$$

where Δr is the distance between the bubble and the pipe wall, $U_z^{gas}, U_r^{gas}, U_z^{Liquid}, and\ U_r^{Liquid}$ are respectively the and radial components for the gas and liquid speed, f_{Gas} and f_{Liquid} are the gas and liquid volume fractions respectively, and R is the radius of the pipe.

At outlet:

$$P_{out} = 101325 \text{ Pa, for } \forall t$$

(8)

At symmetry planes:

$$\frac{\partial U}{\partial \theta} = 0 \text{ , for } \forall (r,t)$$

(9)

where θ it is the angular coordinate.

At pipe walls:

$$r = R \begin{cases} U_z^{Gas} = U_z^{Liquid} = 0.0 \text{ m/s} \text{ for } \forall t \\ U_r^{Gas} = U_r^{Liquid} = 0.0 \text{ m/s} \end{cases}$$

(10)

Thermo Physical Properties and Data for Simulation

The physical properties of liquid and gas used in the simulations are presented in Table 1.

In Table 2 are given other details about the mathematical model and numerical treatment of governing equations.

The cases analyzed in the present work are shown in Table 3.

The Numerical Mesh

The geometry and the dimensions of the pipe used in the present work are shown in the Figure 1. Due to the observed symmetry of the fluid flow in tubes, this study has been realized in an unstructured two-dimensional mesh obtained by using CFX$^{\rightarrow}$ 5.6 software.

Table 1: Physical properties of fluids

Physical Properties	Continuous Phase (oil)	Dispersed Phase (air)
Density (kg/m³)	920	1.185
Dynamic Viscosity (Pa·s)	1.5	0.00001831
Surface Tension (N·m)	0.07	

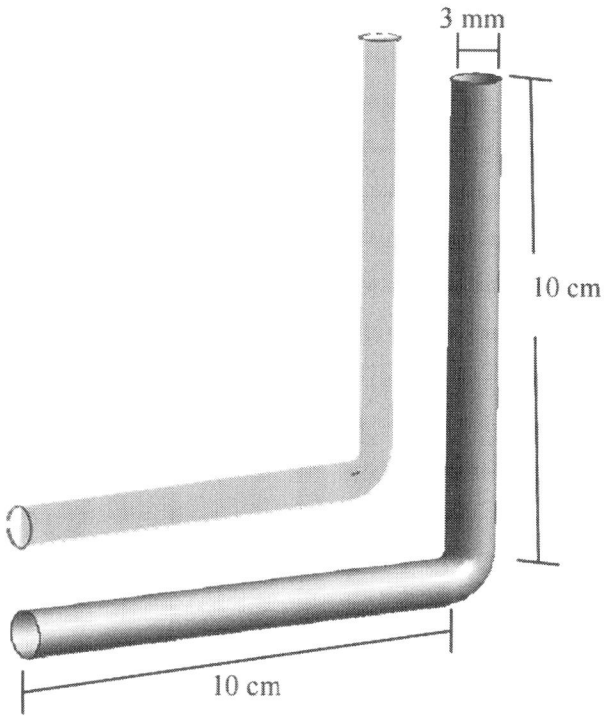

Figure 1: Geometrical shape and dimensions of pipe used in this study.

Table 2: General conditions of the physical problem and numerical treatment

Cases studied	Oil viscosity (Pa·s)	Time of air injection (s)	Air speed (m/s)	Oil speed (m/s)	Air injection nozzle radius (m)	Time of bubble trajectory in pipe (s)
Case 1	1.5	0.02	0.1	0.1	0.001	1.0
Case 2	1.5	0.1	0.1	0.1	0.001	1.0
Case 3	1.5 0.5			0.1	0.1	0.001	1.0
Case 4	0.5	0.5	0.1	0.05	0.0005	2.0
Case 5	1.5	0.5	0.1	0.05	0.0005	2.0
Case 6	2.5	0.5	0.1	0.05	0.0005	2.0
Case 7	5.0	0.5	0.1	0.05	0.0005	2.0

Table 3: Characteristic data of the cases studied

Flow	Biphasic
Flow regime	Transient
Time step (t)	10^{-2}
Fluids used	air (dispersed phase), oil (continuous phase)
Ambient conditions	25°C and 1 atm
Model	Non homogeneous
Inter-phase model transfer	Particle model
Pressure interpolation scheme	Trilinear
Speed interpolation scheme	Trilinear
Influence of wall for oil	No slip condition
Influence of wall to the air	Free slip condition
Mass transfer between phases	None
Advection scheme	High resolution
Drag model coefficient	Grace model
Convergence criterion	Residual mean square (RMS) (10^{-4})
Transient scheme	Second order backward Euler

This mesh constituted of prismatic and pyramidal elements is illustrated in Figure 2.

The behavior of bubble flow in pipe with an angular junction of 90° was investigated numerically using Pentium 4 computers, Core 2 Duo 3.0 GHz, 2048 Mb RAM memory and a hard disk of 120 Gb. The total time of simulation for each case was of approximately of 11 hours.

RESULTS AND DISCUSSION

The results of the numerical study of the fluidodynamics and geometric behavior of air bubbles during the cocurrent air-oil flow inside a pipe with a 90° curvature are presented. The effects of the parameters evaluated are: the air injection time (0.02, 0.1 and 0.5 s) in the horizontal and vertical sections of the pipe, the oil viscosity (0.5 to 5.0 Pa·s) on the format of the bubble and pipe curvature of 90° on the symmetry of the formed bubbles. The details of the cases studied are shown in Table 3.

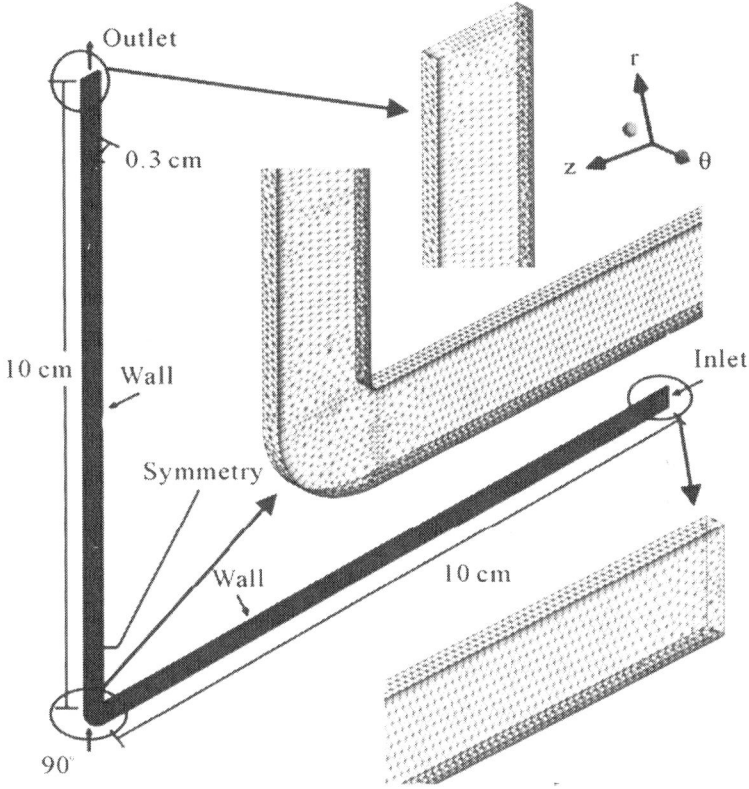

Figure 2: Numerical mesh and pipe boundary.

For representing the air bubble movement in the interior of the pipe, the air phase volumetric fraction distributions have been used. In all the figures, the blue color corresponds to the liquid phase, and other colors show the variation of the air fraction in the gas bubble. To facilitate the visualization and interpretation of the results the minimum and the maximum fractions of the gas were fixed at 0.0 and 0.7, respectively.

Effect of Air Injection Time (Air Concentration)

To analyze the effect of air concentration on the formation and movement of bubbles inside the pipe, certain quantity of gas, through a circular orifice of radius R_{max} = 1 mm was injected into the pipe.

The behavior of the air bubble for the trajectory time of 0.4 s and 1.0 s are shown in Figures 3 and 4 respectively. It can be seen that, for all the three cases, Cases 1, 2 and 3, the diameters of the bubbles formed approach to the diameter of the pipe, and can be considered as Taylor bubbles ([1,2,4,10,12-13,15]).

For Cases 1 and 2, it can be noticed that the bubbles are practically at the same positions of the pipe (see Figure 3). The central parts of the bubbles are approximately at 6.5 cm of the distance. However, this did not happened for Case 3, where the central part of the first bubble is approximately at a distance of 4.5 cm. On further dislocation of the bubbles, in the tube, their frontal parts become more concave. This effect could be due to the friction of the wall contact area on the bubble, which is easily visualized in Case 3.

For Cases 1 and 2, in the horizontal section of the pipe, in the upper back side of the bubble, a formation of gas trail is notice (see Figure 3). This can be due to the release of the micro bubbles. Thus, at the end of 1s of trajectory time, when the Taylor bubble reaches the end of the vertical section of the pipe, its volume is reduced (see Figure 4).

In Case 3, when the amount of air is increased by 25 times with respect to Case 1, formation of more bubbles, dislocating near the upper surface of the tube wall, can be observed (see Figures 3(c) and 4(c)). For this case 3, having 0.4 s of trajectory time, formation of a bigger bubble with three small nuclei, possessing approximately 70% of volumetric air fraction, can be seen. Also, in this case, as the trajectory time of 0.4 s, which is less than the air injection time of 0.5 s, an incomplete release of the bubble at the nozzle mouth is seen and the formation of a micro bubble trail, on the back side of the bubble, is noticed (see Figures 3(c) and 4(c)).

The positions of the first bubbles, in the pipe, for the trajectory time of 1.0 s, shown in Figure 4, are different but consistent with the air injection times. It is to be noted that in Case 3, the air fraction in the second bubble is higher. This phenomenon may be explained due to the reduction of bubble velocity, when it passes through the 90° angle of the pipe. This reduction of the velocity helps in merging of the last bubble with the second one and letting the tail of this bubble disappear. Thus, the size and format profiles of the two bubbles are different, (see Figure 4(c)). The movement of the bubbles in the vertical part of the pipe makes their frontal part neatly spherical.

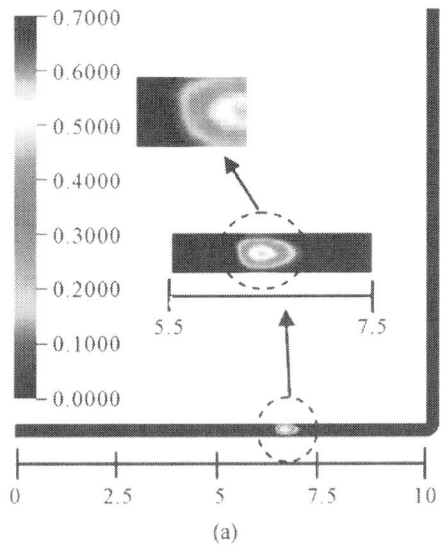

(a)

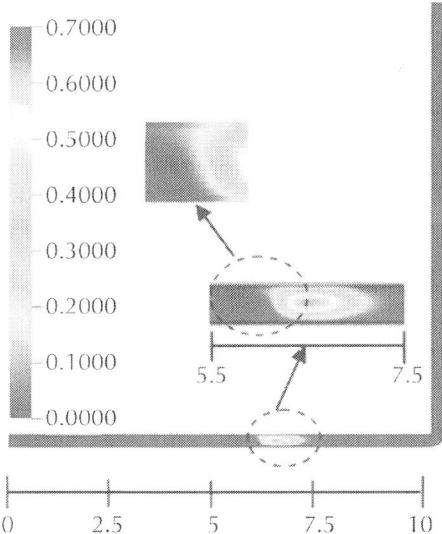

(b)

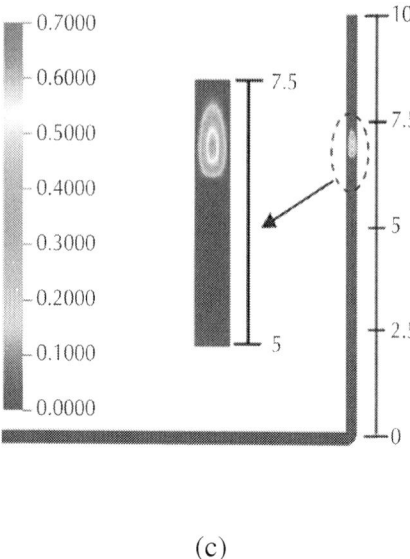

(c)

Figure 3: (a) Volumetric fraction of air in bubble and bubble position at 25°C and t = 0.4 s (Case 1); (b) Volumetric fraction of air in bubble and bubble position at 25°C and t = 0.4 s (Case 2); (c) Volumetric fraction of air in bubble and bubble position at 25°C and t = 0.4 s (Case 3).

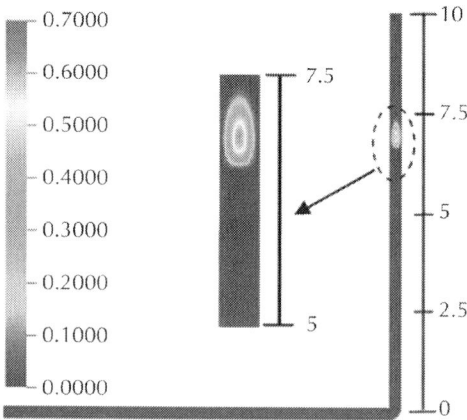

(a)

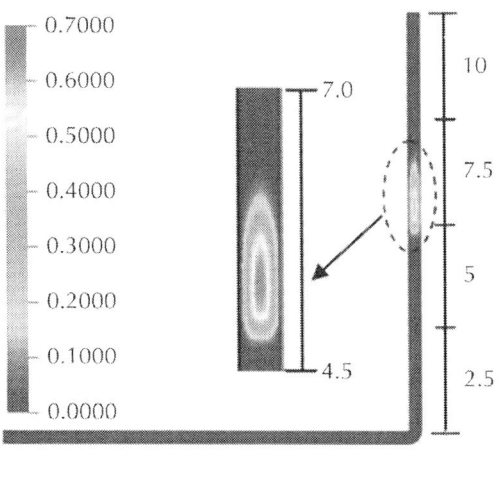

(b)

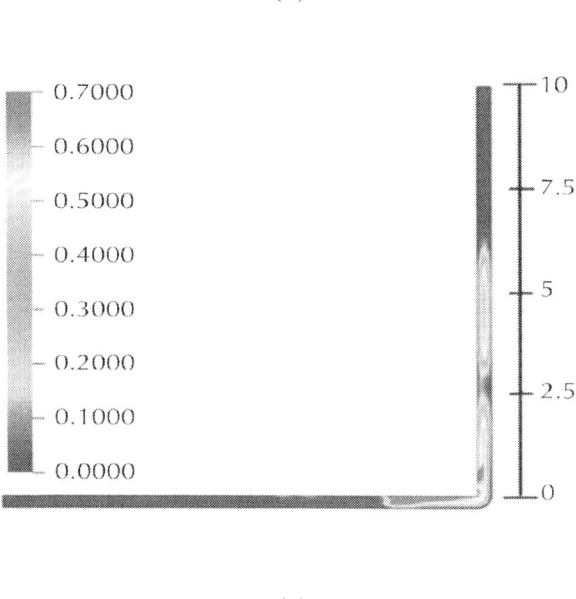

(c)

Figure 4: (a) Volumetric fraction of air in bubble and bubble position at 25°C and t = 1.0 s (Case 1); (b) Volumetric fraction of air in bubble and bubble position at 25°C and t = 1.0 s (Case 2); (c) Volumetric fraction of air in bubble and bubble position at 25°C and t = 1.0 s (Case 3).

This observation is in agreement with the one reported by [10].

Comparing the results, shown in Figures 4(a)-(c), it becomes evident that the air injection time has great influence on the format and the length of the Taylor bubble, which is also mentioned by [10]. These authors commented that as the amount of air injected is increased, the bubble size increases while changing its format from spherical to elliptical due to the buoyancy forces. In all the cases, analyzed here, the asymmetry of the bubble in the horizontal section and symmetry in the vertical section of the pipe has been verified.

Effect of the Oil Viscosity

The oil phase viscosity has a strong effect on the bubble format and has been reported in works of [1,14-16], among others. In Figure 5, the results of the effect of the variation of the viscosity on the bubble, corresponding to Cases 4, 5, 6 and 7, for a 1.4 s of trajectory time and air injection time 0.5 s, are presented. It can be observed that, for all the four viscosities studied, the bubble formats are well defined. For 0.5 Pa·s viscosity case, three bubbles of different sizes are formed, (see Figure 5(a)). For the first bubble a greater enlargement in the upper part of the bubble tail has occurred. The second bubble has a higher air concentration in the central part and the third bubble presents a small gas trail. The number of bubbles formed for viscosities 1.5 Pa·s and 2.5 Pa·s are four each and for 5.0 Pa·s is five (Figures 5(b)-(d)).

These observed effects can be explained due to the variation of oil viscosity because it exerts different resistances to the bubble flow. Thus, the lower the viscosity, the easier the air bubble tends to deform, and the higher the viscosity, the more numbers of bubbles are formed.

Effect of Pipe Angular Geometry

The effect of 90° angular geometry of the pipe on the behavior of the bubble, for Case 5, can be visualized in Figure 6. The trajectory times taken to reach the angular section, for each bubble, were registered. They were: 1.1 s; 1.29 s; 1.5 s and 1.65 s; for first, second, third and fourth bubble respectively.

When the air bubbles arrive at the angular section (the bent), they enter in contact with the vertical wall of the pipe, get deformed according to the bending but do not break. Then, bubbles move vertically upward, while maintaining their radial symmetry. Thus, when the first bubble passes through the pipe bending, it suffers a small deformation, gets elongated and molds itself to the angular geometry of the pipe. Due to superficial tension between the bubble and the fluid and lower fluid velocity, the bubble does not break. It regains its format in the vertical section of the pipe.

This behavior of the bubble is in agreement with the results reported by [12]. According to these authors, when the speed of the continuous fluid is equal of the dispersed fluid, the format of the bubble is maintained during their flow in horizontal and vertical sections of the pipe.

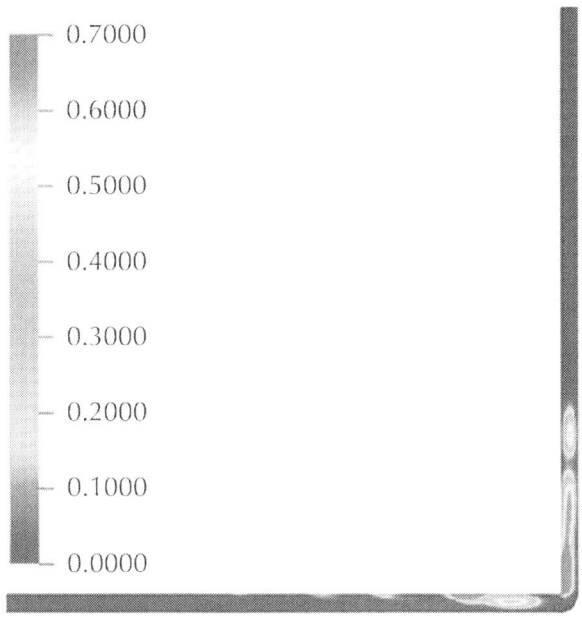

(a)

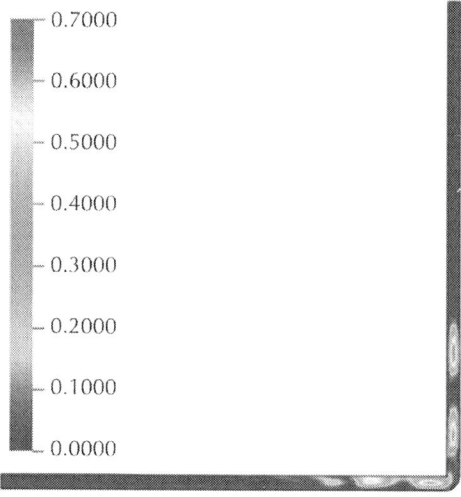

(b)

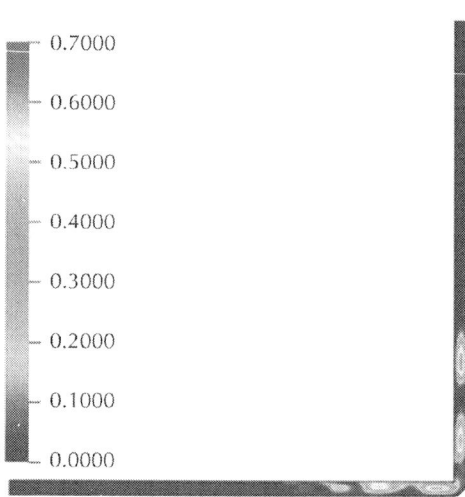

(c)

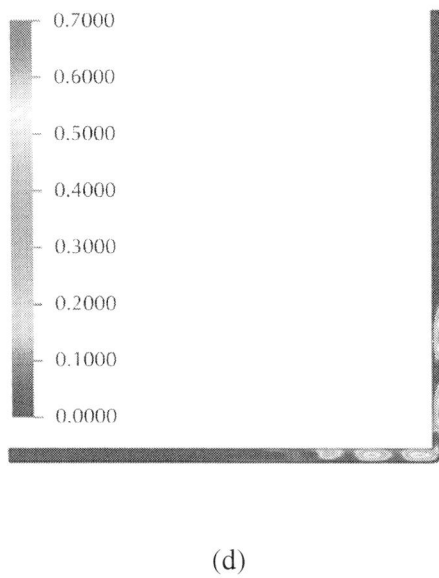

(d)

Figure 5: (a) Volumetric fraction of air in bubble at 25°C, t = 1.4 s and for viscosity of 0.5 Pa·s; (b) Volumetric fraction of air in bubble at 25°C, t = 1.4 s and for viscosity of 1.5 Pa·s; (c) Volumetric fraction of air in bubble at 25°C, t = 1.4 s and for viscosity of 2.5 Pa·s; (d) Volumetric fraction of air in bubble at 25°C, t = 1.4 s and for viscosity of 5.0 Pa·s.

From the volumetric air fraction point of view the first and second bubbles have between 40% to 70% of the gas fraction, Figure 6(a). However, the second bubble is adhered to the upper side of the horizontal section of the pipe due to the buoyancy force and has less concentration of air in its central region. The volumetric fraction of the air in the last bubble is reduced to about 40%. During the trajectory of the bubbles passing through the 90° angular pipe, the concentration of the air reduces from first to last bubble but air fraction in the central region is maintained practically constant, (Figures 6(a)-(d)).

CONCLUSIONS

Based on the results of numerical simulation of the unsteady fluid dynamics of gas bubble flow in 90° curved pipes, following general conclusions can be made.

The format and the length of the Taylor bubble have dependence on the duration of the gas injection time. Higher is the gas injection time, larger is the length of the bubble and more are the bubbles formed.

The air bubble flow, in the pipe, tends to be asymmetrical with reference to the central axis of horizontal section, because of the lower fluid density, and is symmetrical in vertical section, due to the buoyancy effect.

At lower fluid viscosity, the air bubble deforms easily and at higher viscosity more number of bubbles is formed.

The bubble format is molded to the angular geometry of the pipe, while passing through the 90° section, and is regained to the original form in the vertical section, without breaking. It happens due to the superficial tension and lower velocity of the fluids.

ACKNOWLEDGEMENTS

The authors would like to give thanks to CAPES, CNPq, FINEP, PETROBRAS, RPCMOD and JBR Engen- haria LTDA for providing financial support.

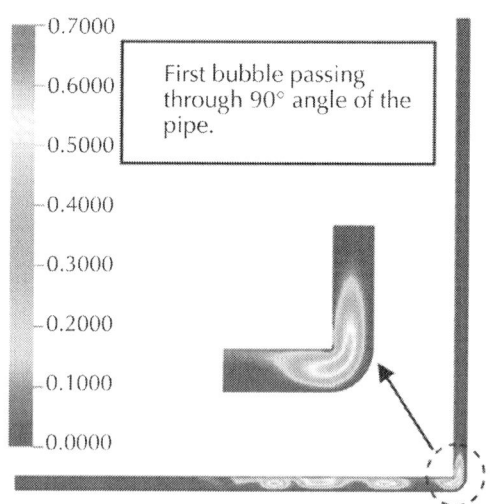

(a)

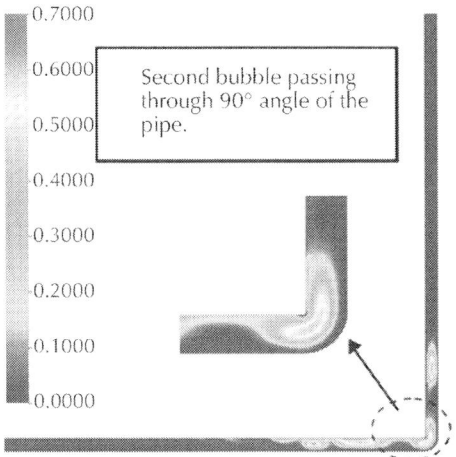

(b)

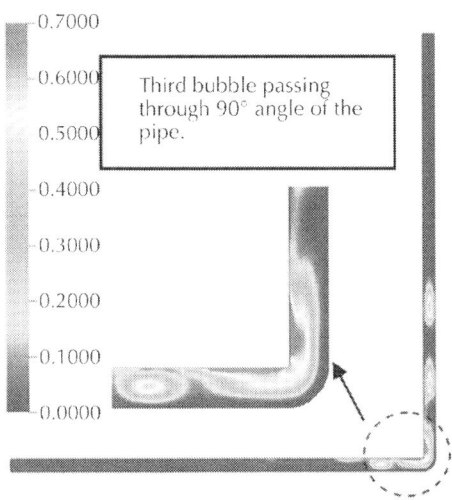

(c)

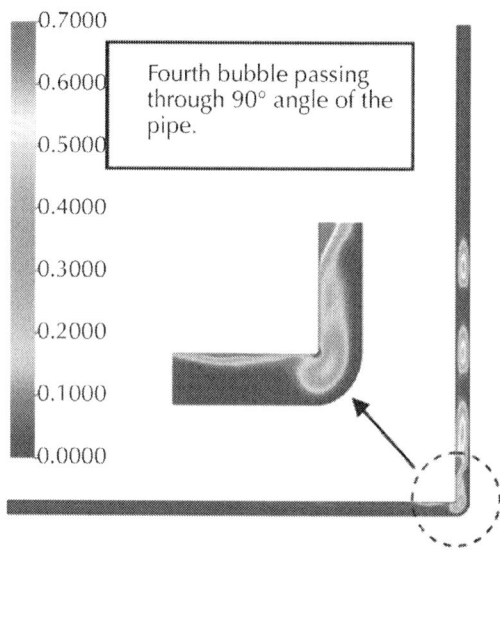

(d)

Figure 6. Volumetric fraction of air in bubble at 25°C and t = 0.5 s , for Case 4 at different trajectory time: (a) 1.1 s; (b) 1.29 s; (c) 1.5 s and (d) 1.65 s.

REFERENCES

1.   J. D. Bugg, K. Mack and K. S. Rezkallah, "A numerical Model of Taylor Bubbles Rising through Stagnant Liquids in Vertical Tubes," International Journal of Multiphase Flow, Vol. 24, No. 2, 1998, pp. 271-281. doi:10.1016/S0301-9322(97)00047-5

2.   T. Cheng and T. Lin, "Characteristics of Gas-Liquid Two-Phase Flow in Small Diameter Inclined Tubes," Chemical Engineering Science, Vol. 56, No. 21-22, 2001, pp. 6393-6398.doi:10.1016/S0009-2509(01)00251-2

3.   S. Madani, O. Caballina and M. Souhar, "Unsteady Dynamics of Taylor Bubble Rising in Vertical Oscillating tubes," International Journal of Multiphase Flow, Vol. 35, No. 4, 2009, pp. 363-375. doi:10.1016/j.ijmultiphaseflow.2009.01.002

4. T. R. Nigmatulin and F. J. Bonetto, "Shape of Taylor Bubbles in Vertical Tubes," Heat Mass Transfer, Vol. 24, No. 8, 1997, pp. 1177-1185. doi:10.1016/S0735-1933(97)00112-7

5. A. M. F. R. Pinto, M. N. C. Pinheiro and J. B. Campos, "On the Interaction of Taylor Bubbles Rising in TwoPhase Co-Current Slug Flow in Vertical Columns: Turbulent Wakes," Experiments in Fluids, Vol. 31, No. 6, 2001, pp. 644-652. doi:10.1007/s003480100310

6. A. M. F. R. Pinto and J. B. L. M. Campos, "Coalescence of Two Gas Slugs Rising in a Vertical Column of Liquid," Chemical Engineering Science, Vol. 51, No. 1, 1996, pp. 45-54. doi:10.1016/0009-2509(95)00254-5

7. G. H. Abdul-Majeed and T. M. Al-Masha, "A Mechanistic Model for Vertical and Inclined Two-Phase Slug Flow," Journal of Petroleum Science and Engineering, Vol. 27, No. 1-2, 2000, pp. 59-67. doi:10.1016/S0920-4105(00)00047-4

8. E. Al-Safran, "Investigation and Prediction of Slug Frequency in Gas/Liquid Horizontal Pipe Flow," Journal of Petroleum Science and Engineering, Vol. 69, No. 1-2, 2009, pp. 143-155. doi:10.1016/j.petrol.2009.08.009

9. G. Bercic and A. Pintar, "The Role of Gas Bubbles and Liquid Slug Lengths on Mass Transport in the Taylor Flow through Capillaries," Chemical Engineering Science, Vol. 52, No. 21-22, 1997, pp. 3709-3719. doi:10.1016/S0009-2509(97)00217-0

10. Q. C. Bi and T. S. Zhao, "Taylor Bubbles in Miniaturized Circular and Noncircular Channels," International Journal of Multiphase Flow, Vol. 27, No. 3, 2001, pp. 561- 570.doi:10.1016/S0301-9322(00)00027-6

11. R. Clift, J. R. Grace and M. E. Weber, "Bubbles, Drops and Particles," Academic Press, California, 1978.

12. D. Qian and A. Lawal, "Numerical Study on Gas and Liquid Slugs for Taylor Flow in a T-Junction Microchannel," Chemical Engineering Science, Vol. 61, No. 23, 2006, pp. 7609-7625. doi:10.1016/j.ces.2006.08.073

13. W. Salman, A. E. Gavriilidis and P. Angeli, "A Model for Predicting Axial Mixing During Gas-Liquid Taylor Flow in Microchannels at Low Boden-Stein Numbers," Chemical Engineering Science, Vol. 101, No. 1-3, 2004, pp. 391-396.doi:10.1016/j.cej.2003.10.027

14. D. Zheng, X. He and D. Che, "CFD Simulations of Hydrodynamic Characteristics in a Gas-Liquid Vertical Upward Slug Flow," International Journal of Heat Mass Transfer, Vol. 50, No. 21-22, 2007, pp. 4151-4165. doi:10.1016/j.ijheatmasstransfer.2007.02.041

15. E. T. White and R. H. Beardmore, "The Velocity of Rise of Single Cylindrical Air Bubbles through Liquid Contained in Vertical Tubes," Chemical Engineering Science, Vol. 17, No. 5, 1962, pp. 351-361. doi:10.1016/0009-2509(62)80036-0

16. ANSYS CFX, "User Manual Theory," USA, 2006.

Chapter 4

Computer-aided Process Engineering in Oil and Gas Production

Michael Nikolaou

Chemical & Biomolecular Engineering Department, University Houston, Houston, TX 77204-4004, United States

ABSTRACT

As oil and gas production moves to ever more challenging areas, increased use of technology becomes increasingly important. One technology that can contribute towards achieving safe and economic production from such areas is computer-aided process engineering. The use of such technology is outlined for two (of many) instances: Managed pressure drilling (MPD) for deep-offshore applications and natural gas production from shales. A broader picture of the importance of computer-aided process engineering for energy is briefly touched upon.

INTRODUCTION

It was less than a couple of years ago when the Macondo well accident in the Gulf of Mexico brought to the forefront the technological challenges and risks – for human life and the environment – faced by the oil and gas industry. At the core of the accident was undetected penetration of hydrocarbons from the rock formation into the well being drilled and subsequent loss of well control (BP, 2010). While the term control in the previous sentence has a different flavor from what is usually understood in the chemical industry (where process control typically refers to the use of automatic feedback for rejection of external disturbances and maintenance of process variables at their setpoints) well control also bears similarity to chemical process control – particularly of the advanced constrained control variety – in that the main objective of well control is to maintain the pressure in a well within well prescribed limits, for safety and performance reasons. To achieve this objective, well control technology has traditionally relied on a combination of well prescribed (no pun intended) work flows and the use of appropriate equipment In that respect, well control bears similarities to aerospace control as well, where automation is fairly advanced with commensurate care for maintenance of situation awareness by human operators of automated systems (Endsley & Garland, 2000). In fact, in the perennial debate whether more or less automation is beneficial (in view of the risks and benefits associated with human involvement in controlling engineered systems) a crucial realization is that automation can be highly beneficial, as long as humans maintain situation awareness (Norman, 1990) and retain critical manual skills (Wiener & Curry, 1980), so that they can take appropriate action in situations for which automation is unprepared. Such skills are valuable in both routine and abnormal situation management (Cochran, Miller, & Bullemer, 1996). Of the many lessons learned in the Macondo well accident, the lesson on the importance of human decisions as well as on the interaction between humans and machines is valuable.

Based on the preceding discussion it is fair to say that automatic control of hydrocarbon well drilling operations bears certain similarities to both chemical process and aerospace control, but it also has its own intricacies and challenges. While control of hydrocarbon well drilling can certainly draw from these disciplines, it certainly poses different enough problems to warrant research on such problems, a task for which chemical engineers have the background to make useful contributions. In fact, as will be outlined below, concepts such as multi-level control and real-time optimization, well established in the chemical industry, can be directly transferred to the oil and gas industry.

A related story can be told about another development that has had a major impact on projections for our energy future, namely the very recent development of unconventional natural gas resources, particularly shale gas. While the term unconventional, by suggesting what such resources are not, provides little insight into what such resources are, the notion of shale gas refers to natural gas trapped in rock of very low porosity (2% or less) and permeability (0.1 to 0.0001 md or even less). Therefore, if a well were drilled into such rock, the gas would take an extremely long time to reach the well, and production would be so low that any economic extraction of such gas from the ground would be infeasible. What has made the recovery of shale gas economically attractive is the extensive use of two crucial technologies: directional drilling (Economides, Watters, & Dunn-Norman, 1994) and hydraulic fracturing (Economides & Nolte, 2000). Identified along with 3D-seismic imaging as the most crucial recent technologies for oil and gas exploration and production (Economides & Nikolaou, 2011), directional drilling and extensive hydraulic fracturing have resulted in dramatic increase of the recoverable natural gas reserves for the US, where shale gas production was pioneered in the Barnett Shale around 2005. The economic implications of this development are significant, as natural gas prices in North America have stayed significantly lower than elsewhere in the world and multi-billion dollar LNG regasification terminals built in the US just before the emergence of shale gas production suddenly became redundant in a market that no longer needed LNG imports and could even afford exports. To appreciate the quantities of shale gas, suffice it to say that proved natural gas reserves in the US totaled about 175 Tcf in 1998; in 2009, after about 250 Tcf of production in the intervening years (around 23 Tcf/year on the average), US proved gas reserves rose to 285 Tcf.

It is now becoming evident that large quantities of recoverable shale gas can be expected in many other areas outside the US, e.g. China (Wang & Wang, 2011), Poland, the UK, and others, for currently estimated global reserves of about 16,000 Tcf, a figure that may change in the future. Developing a resource of such magnitude efficiently poses both engineering and environmental challenges (Moridis et al., 2011b and Sakmar, 2011).

The rest of the paper summarizes some recent work on automated well control and on shale gas resource development. The objective is to delineate major issues along with some specifics that help elucidate the bigger picture.

DRILLING HYDROCARBON WELLS

Drilling hydrocarbon wells is based on a simple principle: A rotating drill bit at the bottom of a drillstring (a long string of thread-joined pieces of pipe and tools suspended from a derrick) creates a hole into a rock formation. The rock cuttings are transferred from the bottom of the drilled hole to the surface by a circulating fluid (drilling "mud"), pumped from the surface to the bottom through the drill pipe and back to the surface through the annulus between the drill pipe and the well walls. The drill bit rotation is provided by drill pipe rotation or, for horizontal wells, by a bottomhole mud motor. The drilling mud serves multiple tasks, such as lubrication, protection of well walls from collapse or damage, and containment of hydrocarbons within the reservoir during drilling, to prevent a blow-out. As mentioned in the Introduction, the importance of failing to maintain wellbore pressure at the right level and to ensure that fluids from the reservoir do not enter the well became painfully familiar to the general public after the Macondo well accident and madeblow-out preventer a household term (Fig. 1).

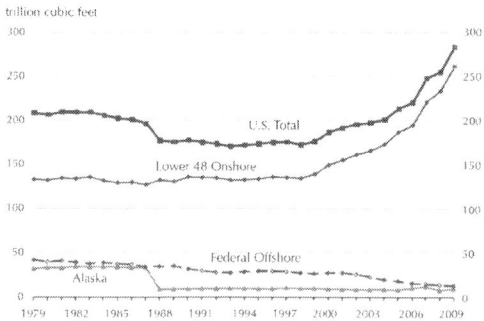

Figure 1: US wet natural gas proved reserves (Energy Information Administration, 2011).

Horizontal and multi-lateral well drilling started in the 1980s, and is now used routinely. Even though technically complex and costly, it offers distinct economic benefits, such as improved contact area with the reservoir (translated into improved hydrocarbon recovery – a particularly important factor for shale gas production), and a convenient- or single-entry point for exploitation of an entire reservoir (a crucial factor for offshore production, where the cost of a platform easily reaches billions of dollars).

MANAGED-PRESSURE DRILLING (MPD)

Maintaining the health of a drilling system entails several concerns such as mechanical integrity and resiliency, vibration control (Spanos, Chevallier, Politis, & Payne, 2003), weight-on-bit control (Nikolaou, Misra, Tam, & Bailey, 2005), drilling fluid flow and consistency, management of surface facilities (such as pumps, mixers, and storage tanks), and, crucially, pressure management in the borehole (Godhavn, 2009). This is a formidable challenge, given that a drillstring traverses several thousand feet into a rock formation, after going through several thousand feet of sea water for offshore applications. To enhance pressure control flexibility, efficiency, and safety, a number of drilling technologies, collectively known as managed pressure drilling (MPD) (Hannegan, 2007), have emerged as a powerful proposition for precise control of wellbore pressure. The primary goal of MPD is to keep wellbore pressure within constraints (Fig. 2). To accomplish this, MPD typically employs a closed, pressurized mud circulation system, along with additional pumps, valves and chokes, in contrast to conventional systems where circulating mud is returned through an open line at atmospheric pressure. Because MPD treats the mud circulation system as a closed vessel, rather than as an open system, it offers higher flexibility and precision than traditional pressure adjustment based on mud weight and mud pump rate adjustments alone. However, this also generates operability challenges that have to be addressed before widespread acceptance of the technology (Rehm, Schubert, Haghshenas, Paknejad, & Hughes, 2008). These challenges emanate, in no small part, from (often non-trivial) interactions among different pieces of equipment and the need for coordinated control of their operation. Work flows that are well accepted in industry (and mostly based on human intervention) have to be adapted and further developed for use in MPD. More importantly, work flows based mostly on humans reach their limits when applied to MPD, given the limited capability of humans to handle reliably situations that involve several interacting variables simultaneously. Inability to handle such situations satisfactorily would result in significant economic loss through loss of productive time and could pose serious safety threats.

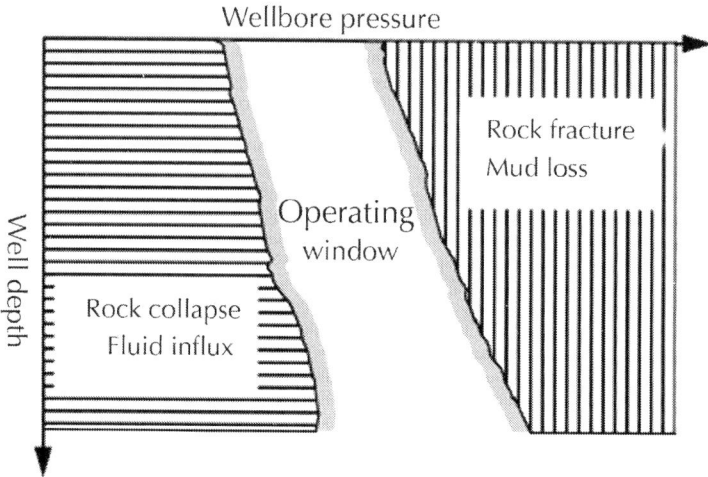

Figure 2: Operating window for wellbore pressure in the annulus between the drill pipe and well wall. If pressure violates its operating bounds, the rock may collapse, fluids from the reservoir may flow into the well (a "kick") possibly causing an accident, the rock may be fractured and/or drilling mud may be lost into the pores of the rock formation.

A solution is the use of enabling automation tools (Breyholtz, Nygaard, & Nikolaou, 2010). Such tools can integrate MPD-related activities, allowing humans to concentrate on higher-level decisions, while leaving the reliable execution of lower level decisions in an increasingly automated fashion. Separation of control tasks at multiple levels, separated mainly on time-scale, is crucial for such a critical automated system to work safely (Breyholtz, Nygaard, Siahaan, & Nikolaou, 2010). It is also important to distinguish how varying degrees of automation affect situation awareness and workload on human operators during the execution of dynamic control tasks (Endsley &

Kaber, 1999). In such situations there are generally four generic functions that can be performed either by human or computer: (1) monitoring: scanning displays to perceive system status; (2) generating: formulating options or strategies for achieving goals; (3) selecting: deciding on a particular option or strategy; and (4) implementing: carrying out the chosen option. Depending on the fraction of the preceding functions assigned to the human or computer, several degrees of automation can be established, as shown, for example, in Table 1 (Endsley & Kaber, 1999).

Table 1: Degrees of automation

Degree of automation	Function
Manual control	Human performs all tasks (e.g. manual steering)
Computer support	Computer aids human in decision making and implementation (e.g. cruise-control)
Consensual automation	Computer implements control with human consent
Monitored automation	Computer implements control unless vetoed by human (e.g. switch to manual mode)
Full automation	No human interaction (e.g. "refrigerator mode")

Although automation has proven its reliability, safety, and ability to outperform humans in the chemical industry and its potential in other areas (e.g. traffic management (Godhavn, Lauvdal, & Egeland, 1996) and aerospace control (Maciejowski & Jones, 2003)) MPD automation has been limited in the field, but its potential is growing (Godhavn et al., 2011, Godhavn, 2009 and Thorogood et al., 2009). The driver is the need for improved safety and lower cost, particularly as oil and gas production moves to challenging offshore areas.

A glimpse of what can be achieved by MPD automation is given for the prototypical model predictive control (MPC) system in Fig. 3 (Breyholtz, Nygaard, & Nikolaou, 2010). In this system, the objective is to maintain bottom hole pressure (BHP) close to a desired value and within safe bounds. Real-time measurements of BHP are available. At a certain point in time, the drillstring is first lifted, thus creating a void in the well that is filled with additional drilling fluid. After staying at this position for 10 min, the drillstring is then lowered to its previous position, now displacing drilling fluid. Following standard procedures, a human operator would concentrate on the position of the hook from which the drillstring is suspended, while keeping the flow rates of the main and subsea pumps constant. In contrast, a multivariable control system can clearly coordinate the flow rates of both pumps in such a way that pressure swings are reduced.

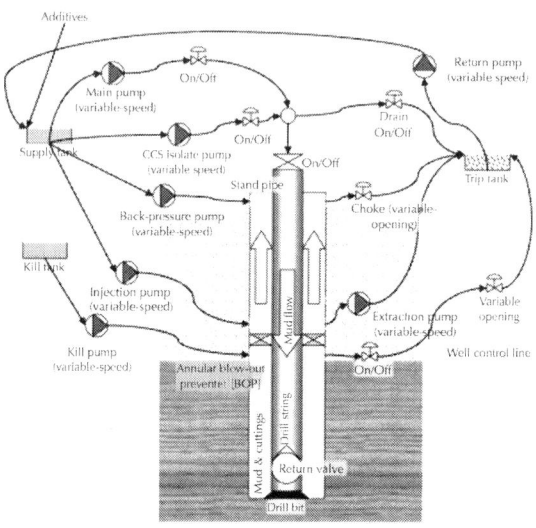

Figure 3: Generic MPD schematic. The similarity between this upstream system and typical downstream chemical plant or refinery units is obvious, and makes the case that process systems engineering and oil and gas production can both contribute to and benefit from each other's experiences.

Typical simulation results are shown in Fig. 4, Fig. 5 and Fig. 6. It is not surprising that with the additional freedom to adjust pump flow rates, BHP variation is reduced significantly. This makes it easier to maintain BHP within safe bounds, as delineated in Fig. 2. A few points should be emphasized here. The proposition to adjust pump flow rates, an innocent proposition for a control engineer, is all but certain to be met with scepticism by human operators. Therefore, to help increase the chance of acceptance of this new idea, the underlying hardware and related algorithms should be as robust as possible, even at the cost of inferior nominal performance. In that respect, the choice of

MPC was based on simple engineering judgement: All that is needed here is a demonstrably sensible multivariable control approach that ensures good coordination of multiple variables while observing crucial constraints. (A more detailed study on various control schemes has been conducted by Carlsen, Nygaard, and Nikolaou (2012).) In fact, involvement of and feedback from the intended final users during the development of related technology would be highly beneficial. At the same time, there should be firm commitment from management to support the installation, commissioning, operation, and further development of such new technology. Given the rich history of MPC, one could reasonably expect that many ideas from both practice and literature would be applicable, and that common themes suitable for fundamental investigation would emerge.

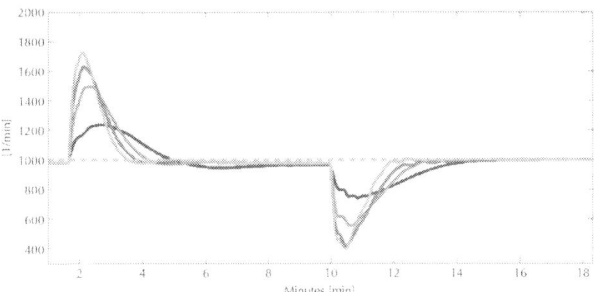

Figure 4: Flow rate from the main mud pump under the standard procedure (dash-dotted line) and multivariable control for four different tunings (continuous lines).

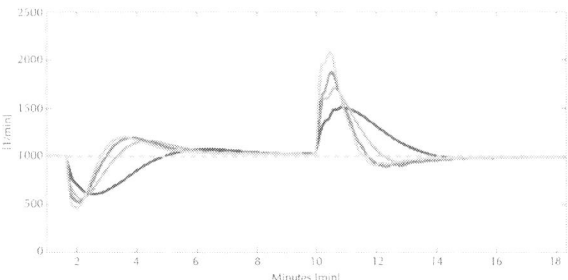

Figure 5: Flow rate from the subsea mud pump under the standard procedure (dash-dotted line) and multivariable control for four different tunings (continuous lines).

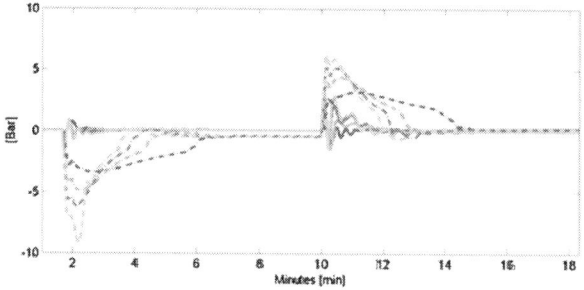

Figure 6: Variation in bottom hole pressure (BHP) while moving the drill-string in and out of the borehole under the standard procedure (dash-dotted line) and multivariable control for four different tunings (continuous lines). The ability of multivariable control to maintain BHP within tighter bounds is evident.

SHALE GAS DEVELOPMENT

If drilling a hydrocarbon well is a conceptually utterly simple concept in principle, albeit extremely complicated in practice, increasing shale gas production through extensively fractured horizontal increases conceptual difficulties only mildly, as indicated in Fig. 7. Of course, the associated practical challenges increase significantly, to the extent that systematic design of shale gas field development systems stands to benefit from the availability of tools that aid engineers in a variety of decision making tasks. Such decisions have a significant effect on both production rate and recovery, as well as on economics.

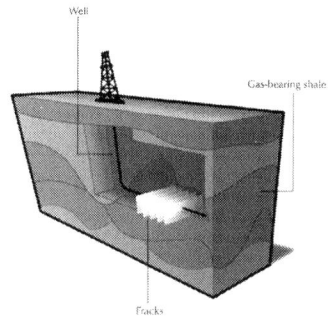

Figure 7: Extensively fractured horizontal well for development of a shale gas field.

There are many questions that need to be answered when designing a fracture treatment, such as the following:

- How many fractures to place?
- What should be the fracture spacing?
- What kind of fracturing fluids must be used and in what quantity?
- What should be the time and rate of injection for each stage of well completion?

There have been many attempts in the recent years to address the above questions. A unified approach to fracture design that produces physically optimal designs given an amount of proppant has been proposed by Economides and co-workers (Economides, Oligney, & Valkó, 2002) for conventional oil and gas resources and has been recently extended to unconventional (low-permeability) resources (Bhattacharya, Nikolaou, & Economides, 2012). An approach that includes economics is summarized in Bhattacharya and Nikolaou (2011). The methodology proposed in the same reference relies on using best estimates of reservoir properties, design objectives, and design constraints to suggest an optimal design that maximizes net present value (NPV, Fig. 8). This methodology is part of a broader effort (SeTES project) to build tools that can aid shale gas field development engineers who are continuously faced with the task of decision making using a variety of data from multiple sources that can easily make the burden overwhelming (Moridis, Reagan, et al., 2011). SeTES is a software system that (a) can incorporate evolving databases involving a variety of types and amounts of relevant data (geological, geophysical, geomechanical, stimulation, petrophysical, reservoir, production) originating from unconventional gas reservoirs, i.e., tight sands, shales or coalbeds, (b) can continuously update its built-in pubic database and refine the underlying decision-making metrics and process, (c) can make recommendations about well location, orientation, stimulation, design and operation, (d) offers predictions of the performance of proposed wells (and quantitative estimates of the corresponding uncertainty), and (e) permits the analysis of data from installed wells for parameter estimation and continuous expansion of the public database.

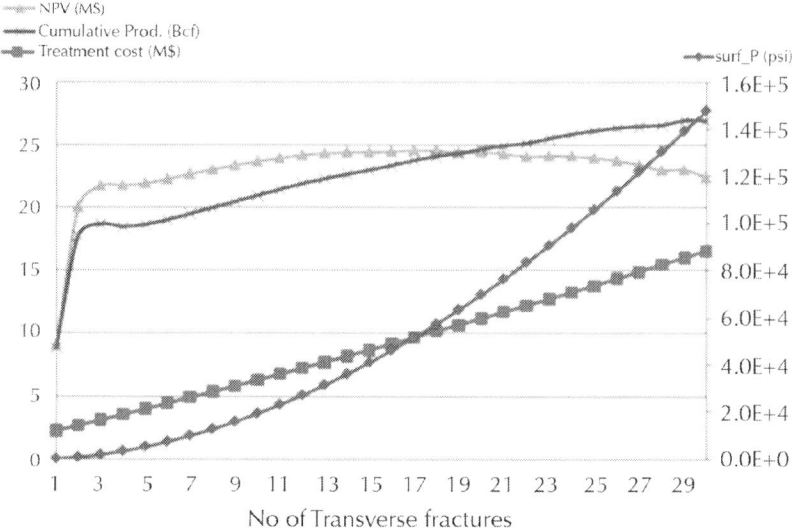

Figure 8: Trade-off between number of fractures and cumulative gas production vs. NPV for a shale gas well (Bhattacharya & Nikolaou, 2011). NPV is initially an increasing function of the number of fractures, until the point of diminishing returns is reached, after which additional gas production does not offset the additional cost of more fractures.

AUTOMATIC CONTROL CHALLENGES

While the preceding two sections outlined two areas where computer-aided process engineering can play an important role, it is also worth discussing here the potential of automatic control for impacting the oil and gas production industry in general, particularly in view of control-related challenges.

As has been mentioned already, there is a solid technological background on which to build automatic control solutions for oil and gas. Certain features of automatic control warrant particular attention, as has been discussed in a recent IFAC Workshop on the subject (Multiple authors, 2012). Some of these features are discussed below.

Steady State and Dynamic Modeling

Drilling (Economides, Watters, & Dunn-Norman, 1998) or producing (Economides et al., 1994) from a hydrocarbon well is, by nature, an unsteady-state operation, although it can be fairly well approximated by pseudo-steady state segments, relatively to pertinent time scales. In addition, it can be argued that no two hydrocarbon wells are the same, and every single well change with time, as it is being drilled or produces. Heraclitus' old dictum "you could not enter the same river twice" (Guthrie, 1997) seems particularly appropriate here. The reason for this variability is that a chemical plant is built to specifications, whereas a hydrocarbon reservoir is already built by nature with conformance to specifications, and knowledge about it is continuously accumulated during exploration, drilling, and production. Reservoir models are frequently used in controlling the drilling and production operations. Such models naturally correspond to different levels of detail, as they are used to make predictions for multiple years (Aziz & Settari, 1979) or days/weeks (Lee, Rollins, & Spivey, 2003) ahead, and use such predictions for corresponding decisions (Nikolaou et al., 2006 and Awasthi et al., 2008). Identification of such models may rely either on existing data - an approach called history-matching - or on experimental data, through well testing (Dake, 1994).

Disturbances and Uncertainty

Multiple disturbances are ever present in drilling and production operations. In fact, such disturbances are so prevalent and potentially catastrophic, that carefully prescribed work flows exist to mitigate detrimental effects and avoid accidents. For example, the boundary of the pressure envelop shown in Fig. 2 is never known exactly, and the possibility exists for fluids to migrate from the reservoir into the well during drilling, due to reservoir pressure being higher than borehole pressure. The so-called Driller's Method or Wait-and-Weight Method are well known manual work flows that can be used to safely handle such a situation (called a "kick") (Grace, 2003). Automation can help make such work flows safer (Godhavn, 2009, Gravdal et al., 2010 and Carlsen et al., 2012).

Nonlinearity

While nonlinear behavior is the norm in models describing oil and gas operations, linear models are often adequate (e.g. Awasthi et al., 2008). However, there are important situations where nonlinearity is significant. Long-term predictions using a reservoir simulator and computation of optimal production profiles is a classic case (Sarma, Durlofsky, Aziz, & Chen, 2006). Another example is the strongly nonlinear relationship between weight-on-bit (WOB) and rate of penetration (ROP) when drilling a well. It has long been known (Dupriest & Koederitz, 2005) that increasing WOB increases ROP up to the founder point, beyond which ROP decreases. This behavior creates a transition from open-loop stable to unstable behavior (Nikolaou et al., 2005) which creates interesting control problems when ROP is at its maximum (Awasthi, 2008).

CLOSING THOUGHTS

Although not the only or even the most important technology to impact the oil and gas industry, computer-aided process engineering can play an important role in future developments of the industry, as computer-based tools keep becoming increasingly powerful. Only a glimpse of recent developments in two areas was presented: managed pressure drilling (MPD) control and shale gas development. In fact, the potential of computer and communications-based solutions has been widely recognized in the industry, as manifest by the recent proliferation of terms such as intelligent fields or similar to denote the extensive use of computers and communications for integrated, remote, and heavily computer-assisted work flows in drilling and production operations. In many respects, the development of such technologies is arguably only a matter of "just doing it", and adoption of such technologies in the field is constrained by other factors (Daneshy & Donnelly, 2004). In other respects, however, formidable challenges remain in practice, since implementation of the right technologies, suitably adapted and further developed for the field is not trivial. Such challenges will keep emerging as oil and gas production is moving to places which only a short time ago would have been thought infeasible (e.g., deeper offshore, tar sands, gas and oil shales (unconventional resources,

Stark, Chew, & Fryklund, 2007), the arctic (Aggarwal & 'Souza, 2011), natural gas hydrates (Moridis, Collett, et al., 2011), and others).

Of course, the bigger challenge is how far into the future the ultimate decline of oil and gas can be pushed through technology, while transitioning to alternative energy sources (Smalley, 2005). The time-scale ofultimate in the preceding sentence is highly uncertain. It is telling that projections (not to be confused withpredictions) over the next 25 years portray an increase in the global use of all fossil fuels, with their percentage in the total energy make-up almost intact (Fig. 9). It is actually ironic that coal, the cheapest fossil fuel but largest greenhouse gas (GHG) producer has increased the fastest since early 2000 (mainly because of China's frenetic economic growth). The non-interchangeability of different forms of energy should also not be missed. For example, US transportation currently runs almost entirely on oil (with corn-based ethanol recently making a small dent) whereas practically no oil is used for electricity generation. It is also interesting that switching from one fossil fuel (coal) to another (natural gas) could result in significant reduction of GHG emissions, albeit at a non-trivial cost. Finally the role of unforeseen rare events should not be underestimated either, as the Fukushima nuclear disaster in March 2011 exemplified. Interestingly, evolutionary nuclear technologies had been singled out by the NRC (National Research Council, 2009) as one of the two key CO_2 emissions reduction technologies that should be immediately funded for large-scale development (the other being carbon capture and geological storage).

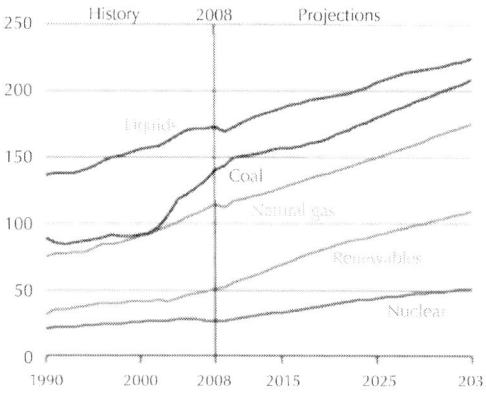

Figure 9: World energy consumption by fuel (Energy Information Administration, 2011).

Regarding future developments, for which, it should be stressed, one can project but hardly predict, a systems approach will be crucial, in that energy systems – from generation to distribution to consumption – are highly integrated and optimizing or controlling one part does not result in global optimum. While the focus of this paper has been computer-aided process engineering for oil and gas, it can be convincingly argued that similar concepts are vital for the broader field of energy, and related contributions from chemical engineers would be both feasible and welcome.

ACKNOWLEDGMENTS

Partial financial support from RPSEA (Research Partnership to secure Energy for America) and IRIS (International Research Institute of Stavanger, Norway) is gratefully acknowledged. Thanks are also due to Tony Rose for preparing Fig. 7.

REFERENCES

1.   Aggarwal, R., & 'Souza, R. D. (2011). Deepwater Arctic – Technical challenges and solutions. In OTC Arctic technology conference Houston, TX, USA.

2.   Awasthi, A. (2008). Intelligent oilfield operations with application to drilling and production of hydrocarbon wells. PhD Thesis, Chemical Engineering, Houston, TX,

3.   University of Houston.

4.   Awasthi, A., Sankaran, S., Nikolaou, M., Saputelli, L., & Mijares, G. (2008). Short-term production optimization by automated adaptive modeling and control. SPE 112239- MS.

5.   Aziz, K., & Settari, A. (1979). Petroleum reservoir simulation. Applied Science Publishers.

6.   Bhattacharya, S., & Nikolaou, M. (2011). Optimal fracture spacing and stimulation design for horizontal wells in unconventional gas reservoirs. In SPE 147622, SPE ATCE Denver, CO.

7.   Bhattacharya, S., Nikolaou, M., & Economides, M. J.(2012). Unified fracture design for very low permeability reservoirs. Journal of Natural Gas Science and Engineering, 9, 184–195.

8. BP. (2010). Deepwater horizon accident investigation report. http://www.bp.com/ liveassets/bp internet/globalbp/globalbp uk english/gom response/STAGING/ local assets/downloads pdfs/ Deepwater Horizon Accident Investigation Report.pdf

9. Breyholtz, Ø., Nygaard, G., & Nikolaou, M. (2010). Automatic control of managed pressure drilling. In American control conference Baltimore, MD.

10. Breyholtz, Ø., Nygaard, G., Siahaan, H., & Nikolaou, M. (2010). Managed pressure drilling: A multi-level control approach. In SPE 128151-MS, SPE intelligent energy conference and exhibition Utrecht, the Netherlands.

11. Carlsen, L. A., Nygaard, G., & Nikolaou, M. (2012). Evaluation of control methods for drilling operations with unexpected gas influx. Journal of Process Control.

12. Cochran, E. L., Miller, C., & Bullemer, P. (1996). Abnormal situation management in petrochemical plants: Can a Pilot's Associate crack crude? In Proceedings of the IEEE 1996 national aerospace and electronics conference, NAECON 1996.

13. Dake, L. P. (1994). The practice of reservoir engineering. Elsevier. Daneshy, A., & Donnelly, J. (2004). A JPT roundtable: The funding and uptake of new upstream technology. Journal of Petroleum Technology, 28–30.

14. Dupriest, F. E., & Koederitz, W. L. (2005). Maximizing drill rates with real-time surveillance of mechanical specific energy. In SPE/ IADC drilling conference, SPE 2194 Amsterdam, Netherlands.

15. Economides, M. J., Hill, A. D., & Ehlig-Economides, C. (1994). Petroleum production systems. Prentice Hall.

16. Economides, M. J., & Nikolaou, M. (2011). Technologies for oil and gas production: Present and future. AIChE Journal, 57(8), 1974–1982.

17. Economides, M. J., & Nolte, K. G. (2000). Reservoir stimulation. John Wiley & Sons.

18. Economides, M. J., Oligney, R. E., & Valkó, P.(2002). Unified fracture design. Orsa Press.

19. Economides, M. J., Watters, L. T., & Dunn-Norman, S. (1998). Petroleum well construction. Wiley.

20. Endsley, M. R., & Garland, D. J. (2000). Situation awareness analysis and measurement. Mahwah, NJ: Lawrence Erlbaum Associates.

21. Endsley, M. R., & Kaber, D. B. (1999). Level of automation eOEects on performance, situation awareness and workload in a dynamic control task. Ergonomics, 42(3), 462–492.

22. Energy Information Administration. (2011). International energy outlook 2011. http://www.eia.gov/forecasts/ieo/index.cfm

23. Godhavn, J.-M. (2009). Control requirements for high-end automatic MPD operations. In SPE/IADC drilling conference and exhibition, SPE 119442 Amsterdam, the Netherlands.

24. Godhavn, J.-M., Lauvdal, T., & Egeland, O. (1996). Hybrid control in sea traffic management systems. Hybrid Systems III Verification and Control.

25. Godhavn, J.-M., Pavlov, A., Kaasa, G.-O., & Rolland, N.-L. (2011). Drilling seeking automatic control solutions. Milano, IT: IFAC World Congress.

26. Grace, R. D. (2003). Blowout and well control handbook. Burlington, MA: Elsevier Science.

27. Gravdal, J. E., Nikolaou, M., Breyholtz, O., & Carlsen, L. A. (2010). Improved kick management during MPD by real-time pore-pressure estimation. SPE Drilling & Completion, 12.

28. Guthrie, W. K. C. (1997). A history of greek philosophy: Volume 1. In The earlier presocratics and the pythagoreans. Cambridge University Press.

29. Hannegan, D. M. (2007). Managed pressure drilling. In SPE 2006–2007 distinguished lecturer series.

30. Lee, J., Rollins, J. B., & Spivey, J. P. (2003). Pressure transient testing. SPE textbook series Society of Petroleum Engineers.

31. Maciejowski, J. M., & Jones, C. N. (2003). MPC fault-tolerant flight control case study: Flight 1862. In IFAC safeprocess conference Washington, DC.

32. Moridis, G. J., Collett, T. S., Pooladi-Darvish, M., Hancock, S., Santamarina, C., Boswell, R., et al. (2011). Challenges, uncertainties, and issues facing gas production from gas-hydrate deposits. SPE Reservoir Evaluation & Engineering, 14(1), 76–112.

33. Moridis, G. J., Reagan, M. T., Santos, R., Boyle, K., Yang, W., Kuzma-Anderson, H., et al. (2011). SeTES: A self-teaching expert system for the analysis, design, and prediction of gas production from unconventional gas resources. In SPE 149485, Canadian unconventional resources conference Calgary, Alberta.

34. Multiple authors. (2012). IFAC workshop on automatic control in offshore oil and gas production Trondheim, Norway, http://www.ifac-oilfield.no/

35. National Research Council. (2009). America's energy future: Technology and transformation: Summary edition. National Academic Press.

36. Nikolaou, M., Cullick, A. S., & Saputelli, L. (2006) Production optimization – A moving-horizon approach In SPE 99358, intelligent energy 2006 Amsterdam.

37. Nikolaou, M., Misra, P., Tam, V. H., & Bailey, A. D., III. (2005). Complexity in semiconductor manufacturing, activity of antimicrobial agents, and drilling of hydrocarbon wells: Common themes and case studies. Computers and Chemical Engineering, 29, 2266–2289.

38. Norman, D. A. (1990) The problem of automation:Inappropriate feedback and interaction, not over-automation. Philosophical Transactions of the Royal Society of London, B Biological Sciences, 327(1241), 585–593.

39. Rehm, B., Schubert, J., Haghshenas, A., Paknejad, A. S., & Hughes, J. (2008). Managed pressure drilling Houston, TX: Gulf Publishing.

40. Sakmar, S. L. (2011). Shale gas development in North America: An overview of the regulatory and environmental challenges facing the industry. In SPE 144279-MS, North American unconventional gas conference and exhibition The Woodlands, TX.

41. Sarma, P., Durlofsky, L., Aziz, K., & Chen, W. (2006) efficient real-time reservoir management using adjoint-based optimal control and model updating. Computational Geosciences, 10(1), 3–36.

42. Smalley, R. E. (2005). Future global energy prosperity: The terawatt challenge. Mrs Bulletin, 30(6), 412–417.

43. Spanos, P., Chevallier, A., Politis, N., & Payne, M. (2003) Oil and gas well drilling: A vibrations perspective. The Shock and Vibration Digest, 35, 85–103.

44. Stark, P., Chew, K., & Fryklund, B. (2007) the role of unconventional hydrocarbon resources in shaping the energy future In International petroleum technology conference Dubai, U.A.E.

45. Thorogood, J. L., Aldred, W. D., Florence, A. F., & Iversen, F. (2009) Drilling automation: Technologies, terminology, and parallels with other industries In SPE 11988, SPE/IADC drilling conference and exhibition Amsterdam, the Netherlands.

46. Wang, X., & Wang, T. (2011) the shale gas potential of China In 142304-MS, SPE production and operations symposium Oklahoma City, OK.

47. Wiener, E. L., & Curry, R. E. (1980) Flight-deck automation: Promises and problems. Ergonomics, 23(10), 995–1011.

Chapter 5

A Multi-Scale Approach to Modelling Spatial and Dynamic Ecological Patterns for Reservoir's Water Quality Management

Edna Cabecinha[a], Martinho Lourenço[b], João Paulo Moura[c], Miguel Ângelo Pardal[d], and João Alexandre Cabral[a]

[a]Laboratory of Applied Ecology, CITAB - Centre for the Research and Technology of Agro-Environment and Biological Sciences

University of Trás-os-Montes e Alto Douro, 5000-911, Vila Real, Portugal

[b]CGUC, Department of Geology, University of Trás-os-Montes e Alto Douro, 5000-911, Vila Real, Portugal

[c]Knowledge Engineering and Decision Support Research Center, Department of Engineering, University of Trás-os-Montes e Alto Douro, Vila Real, Portugal

[d]IMAR (Institute of Marine Research), Department of Zoology, University of Coimbra, 3004-517, Coimbra, Portugal

ABSTRACT

With growing levels of urbanization and agriculture throughout the world, it is increasingly important that both research and management efforts take into account the effects of this widespread landscape alteration and its consequences for natural systems. Freshwater ecosystems, namely reservoirs, are particularly sensitive to land use changes. In this context, modelling can be very useful, for decision support, as an investigative tool to forecast the outcome of various scenarios, to guide current management in order to meet future targets and to develop integrated frameworks for management accordingly to the Water Framework Directive (WFD). The present paper examined the applicability of a holistic Stochastic-Dynamic Methodology (StDM), coupled with a Cellular Automata (CA) model, in capturing how expected changes at land use level will alter the ecological status of lentic ecosystems, namely at physicochemical and biological levels. The methodology was applied to Portuguese reservoirs located in the Douro's basin and illustrated with a series of stochastic-dynamic and spatial outputs taking into account expected scenarios regarding land use changes. Overall, the simulation results are encouraging since they seem to demonstrate the tool reliability in capturing the stochastic environmental dynamics of the selected metrics facing spatial explicit scenarios. The ultimate goal was to couple monitoring assessment and the described modelling techniques to ease management and decision making regarding the practical implementation of the WFD, both at the scale of the reservoir body and at the scale of the respective river watershed dynamics.

INTRODUCTION

The EC Directive 2000/60/EC established a framework for Community action in the field of water policy, commonly known as the Water Framework Directive (WFD), aims to prevent further deterioration and to protect and enhance the status of aquatic ecosystems throughout the European Member States till 2015 (European Union, 2000). In densely populated countries with a myriad of competing uses including housing, transport, agriculture and industrial development, the environmental assessment is pushed to assist with land use planning decisions and projections of 'what if' scenarios at the landscape scale and, consequently, it is necessary to capture the main cause–effect relationships between human activity and ecosystem responses (Bailey et al., 2007). The reservoirs, as other freshwater ecosystems, are mainly sensitive to the effects of large-scale land transformation, namely as a result of urbanization and agriculture intensification, since they receive water and materials from the respective watersheds (Alberti et al., 2007). In this context, any management option must take into account not only the components of the reservoir, but also the human activities within the respective watershed.

Ecological modelling started with Lotka–Volterra and Streeter–Phelps in the 1920s, while the comprehensive use of models in environmental management started in the beginning of the 1970s. Meanwhile many models have been developed and today there are hundreds of ecological models which have been used as tool in research or environmental management (see Jørgensen, 1995, Jørgensen, 1999 and Jørgensen, 2005). For decision support, modelling can be very useful as an investigative tool to forecast the outcome of various scenarios, to guide current management in order to meet future targets and to develop integrated frameworks for management (Malafant and Fordham, 1998 and Schauser et al., 2003). Ecological integrity assessment have been improved by creating dynamic models that simultaneously attempt to capture the structure and the composition in systems affected by long-term environmental disturbances (Jørgensen, 1994, Costanza and Voinov, 2003, Chaloupka, 2002, Cabecinha et al., 2004, Cabecinha et al., 2007, Cabecinha et al., 2009a, Silva-Santos et al., 2006 and Silva-Santos et al., 2008). The application of ecological models can synthesize the pieces of ecological knowledge, emphasizing the need for a holistic view of a certain environmental problem (Jørgensen, 2001, Cabecinha et al., 2004, Cabecinha et al., 2007, Cabecinha et al., 2009a, Silva-Santos et al., 2006 and Silva-Santos et al., 2008).

In this context, a new Stochastic-Dynamic Methodology (StDM) has been developed as a mechanistic understanding of the holistic ecological processes, by using appropriate statistical and dynamic modelling techniques (Santos and Cabral, 2004, Cabecinha et al., 2004, Cabecinha et al., 2007, Cabecinha et al., 2009a, Silva-Santos et al., 2006 and Silva-Santos et al., 2008). This recent research is based on the premise that the general statistical patterns of ecological phenomena are emergent indicia of complex ecological processes that do indeed reflect the operation of universal law-like mechanisms. The StDM is a sequential modelling process developed in order to predict the ecological status of changed ecosystems, from which management strategies can be designed. This methodology was successfully tested in several types of ecological systems, such as mountain running waters (Cabecinha et al., 2004 and Cabecinha et al., 2007), reservoirs (Cabecinha et al., 2009a), mediterranean agroecosystems (Santos and Cabral, 2004 and Cabral et al., 2007), estuaries (Silva-Santos et al., 2006 and Silva-Santos et al., 2008), and for simulating the impact of socio-economic trends on threatened species (Santos et al., 2007).

After the validation procedures carried out in a previous work (Cabecinha et al., 2009a), the present paper examined the applicability of StDM model in capturing how expected changes at land use level will the ecological status, namely at physicochemical and biological levels, in Portuguese reservoirs. Although the model system needs to be dynamic in order to construct a range of different scenarios, to ensure that the results are of use by planners and police makers, they need to be presented also in a spatial format. In this scope, the dynamic model used to produce scenarios for StDM simulations was also coupled with a Cellular Automata (CA) model. CA models can be represented by a set of simple production rules while its outcome may mimic a very complex system (Firebaugh, 1988). Therefore they have been increasingly used to simulate urban sprawl and land use dynamics (Yang et al., 2008). Since Land use change is driven by the interaction in space and time between biophysical and human dimensions, the use of GIS (Geographic Information System) databases can be combined to address complex spatial problems in a transparent and useful way to meet the requirements of legislation, such as the WFD to manage catchments from the field to landscape scale.

The goal of the present work is to demonstrate the potential of these two complementary techniques, the StDM and the CA, in the scope of the practical implementation of the WFD. Therefore, when applied as a multi-scale approach for realistic scenarios, the StDM model can be run at different levels simultaneously taking into account stochastic/random phenomena that characterize the real ecological processes (Cabecinha et al., 2009a). Moreover, this paper discusses the application of an interactive GIS tool to discern the spatial patterns of such scenarios in a visual rather than a mathematical format. The Douro river basin, was used as an exemplificative case study, with a focus in Northeast Portugal. In this context, the reservoir dynamics responsive to changes in catchment inputs of nutrients due to land use changes was tested in distinct expected scenarios, based in the Catchment Plan of the Douro river (INAG, 2008), namely regarding the evaluation of the reduction of pollutants achieved through management measures. The overall goal was to couple monitoring assessment and modelling approaches to ease the management and decision making regarding the requirements the practical implementation of the WFD, both at the scale of the reservoir body and at the scale of the respective river watershed dynamics.

MATERIALS AND METHODS

Study Area

This study was carried out in 11 reservoirs from the Douro river catchment (North of Portugal): Miranda (MRD), Picote (PCT), Bemposta (BMP), Pocinho (PCN), Valeira (VAL), Vilar-Tabuaço (VLR), Régua (RG), Varosa (VRS), Carrapatelo (CRP), Torrão (TR) and Crestuma (CRT) (Fig. 1). The main purpose of all these reservoirs is hydroelectric power generation, although some secondary uses are also common, such as navigation, irrigation, water supply and recreation (Cabecinha et al., 2009b).

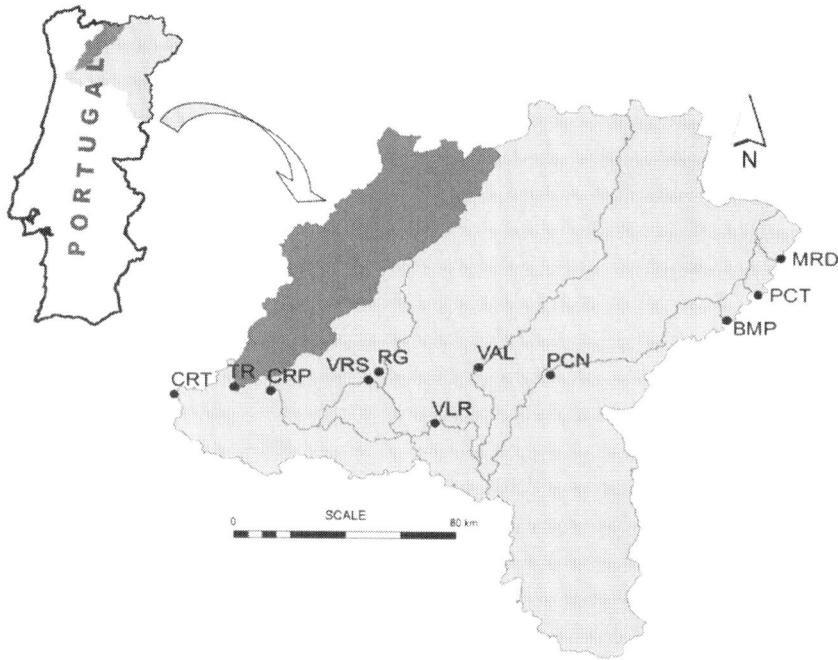

Figure 1: Location of the study area in the Douro river basin with the different watersheds and respective reservoirs, used as data sources in the construction of the StDM model: Crestuma (CRT), Carrapatelo (CRP), Varosa (VRS), Rágua (RG), Vilar-Tabuaço (VLR), Valeira (VAL), Pocinho (PCN), Bemposta (BMP), Picote (PCT) and Miranda (MRD). The data from the Torrão (TR) reservoir (dashed area) was used for scenarios simulation purposes.

The Douro River flows within the largest watershed in the Iberian Peninsula. The flow regime depends not only on climatic conditions but also on hydroelectric power generation needs in Spain and Portugal, as well as by irrigation needs, particularly in Spain. This extensive geographic area represents a wide range in physical and chemical characteristics, soil use and anthropogenic pressure, including both good and poor water quality conditions. Most of the population lives in coastal areas. Therefore, many impacts associated with urbanization are present there, namely water quality problems associated with nutrient enrichment and high biochemical oxygen demand due to industrial effluent discharges, urban development and intensive agriculture.

In general, land use of the Douro basin is dominated by agricultural activities, although the high concentration of industries, mainly transformation industries and mines, is also important (Cabecinha et al., 2009b).

The majority of the study reservoirs are "run-of-river" reservoirs (see Table 1 for details), that more resembles a river than a lake with good mixing and relative high water velocities. The hydraulic retention times in the riverine reservoirs (BMP, CRP, CRT, MRD, PCN, PCT, RG and VAL) are short, with a hydrological stability mostly conditioned by the short-term atmospheric conditions. The remaining reservoirs (VLR, VRS and TR) are explored as "artificial lake reservoir". In this type of reservoirs water storage and release cycles are long, operating on seasonal or multi-year cycles. Besides the removal of suspended sediments and nutrients, these types of reservoirs have major impacts on the discharge and thermal regimes downstream of the reservoirs (Klaver et al., 2007). The main characteristics of the studied reservoirs are presented in Table 1.

Table 1: Codes and mean values for all the variables used in the StDM model construction and simulations

Variables	Code	Bemposta BMP	Carrapatelo CRP	Crestuma CRT	Miranda MRD	Pocinho PCN	Picote PCT	Régua RG	Valeira Val	Vilar VLR	Varosa VRS	Torrão TR
Environmental variables												
Surface water temperature (°C)	Temp	15.6	16.5	16.8	13.3	14.9	16.3	15.6	12.3	15.8	16.2	19.2
Turbidity (NTU)	Turb	1.67	1.69	3.13	10.8	4.97	4.06	4.26	4.85	2.37	3.16	2.29
pH (units)	pH	8.23	7.82	7.70	7.95	8.03	8.13	7.78	7.89	7.74	7.85	7.78
Dissolved Oxygen (mg/L)	DO	7.98	8.29	9.30	9.20	10.8	8.61	10.3	9.53	9.57	8.88	9.29
Hardness (mg CaCo/L)	Hard	168	115	104	181	135	171	127	137	11.1	22.2	23.5
Ammonia-N (mg NH$_4$/L)	NH4	0.180	0.110	0.100	0.270	0.150	0.130	0.140	0.170	0.160	0.960	0.110
Nitrate-N (mg NO$_3$/L)	NO3	5.33	4.77	5.12	7.93	6.05	6.24	7.26	6.90	0.700	3.78	2.60
Phosphorus (mg PO$_4$/L)	PO4	0.270	0.180	0.130	0.250	0.170	0.180	0.090	0.200	0.030	0.270	0.030
Sulfate (mg SO$_4$/L)	SO4	52.7	31.9	30.4	55.4	37.9	51.1	36.3	39.2	3.29	6.74	6.45
Cloride (mg Cl/L)	Cl	19.4	13.2	12.7	17.9	14.6	17.4	12.9	15.2	5.92	11.2	7.64
5-day Biochemical Oxygen Demand (mg O/L)	BDO	2.54	1.40	1.53	2.07	1.85	2.23	1.91	1.84	1.86	3.38	1.44
Total Silicon (mg SiO$_2$/L)	SiO$_2$	1.26	3.14	3.57	3.15	3.29	1.69	4.34	7.28	2.02	8.41	4.94
Total Coliform (N/100 ml)	Tot Colif	215	945	1347	1449	775	507	1869	943.7	199	2613	823
Altitude (m)	Alt	402	71.9	13.2	528.	125	480	73.5	105.2	552	264	65.0

Variable	Abbrev											
Precipitation (mm)	CPRHC	53.4	70.2	90.4	53.9	58.9	53.4	65.0	60.5	79.4	134	122.7
Catchment area (km²)	A	63,850	92,050	92,040	63,100	81,005	63,750	90,800	85,400	370	310	3252
Mean Dam depth (m)	Depth	30.8	16.7	12.9	31.9	15.6	26.9	12.1	11.5	15.7	23.5	20.7
Time of Residence (days)	TimeRes	9.52	5.76	2.24	1.45	2.50	3.27	2.10	3.39	320		13.5
Volume (dam³)	Vol	1,23,267	1,41,966	1,01,563	25,648	94,353	58,946	80,005	92,992	53,567	4526	91,473
Level (m)	LV	400	45.8	12.6	526	107.5	469	72.9	104	544	245	60.2
Biological variables												
Phytoplankton (no. of species)												
Cyanophyta	CN	19.0	15.0	16.0	10.0	18.0	18.0	8.00	8.00	20.0	17.0	22.0
Bacillariophyta	DTM	29.0	28.0	36.0	31.0	30.0	30.0	30.0	26.0	22.0	18.0	24.0
Chlorophyta	CLP	37.0	30.0	41.0	31.0	31.0	33.0	30.0	30.0	22.0	34.0	40.0
Chlorophyll a (mg/l)	Clp_a	0.890	0.610	0.690	0.850	0.760	0.990	0.810	0.770	1.06	1.11	0.770
Soil use												
Artificial Territories (ha)	ART TERT	111	1524	778	33.5	2381	208	1292	1332	141	328	3855
Irrigated crops (ha)	ICROPS	1786	2412	4476	607	32,454	3107	34,812	61,767	3489	1961	22,698
Non-irrigated crops (ha)	NICROPS	0.00	1508	540	0.00	847	0.000	167	286	0.000	259	2026
Vineyard (ha)	VIN	2779	16,056	0.00	27.6	20,136	353	38,908	12,596	614	2258	3845
Orchards (ha)	OCHD	0.000	242	0.00	0.00	16,111	0.000	8204	7427	405	2139	124
Olive grove (ha)	OIV	81.6	171	0.00	0.00	6613	0.000	22,027	18,314	0.000	0.000	219
Grasslands (ha)	GRS	0.800	578	862	0.00	2784	197	2163	1379	0.000	1099	4431

	HTAG										
Heterogeneity agricultural areas (ha)	1741	25,264	26,510	2543	93,437	11,547	1,23,945	1,18,579	6893	7203	73,356
Forest (ha) FRT	1607	17,110	46,514	449	28,559	1147	53,713	34,339	9145	3045	73,462
Shrubs (ha) SRB	1451	32,365	36,099	1048	1,58,413	1706	1,35,861	1,20,535	6517	11,527	99,009
Unproductive areas (ha) UNPRD	0.000	5915	5601	0.000	1843	0.000	10,548	3457	5499	830	10,988
Burned areas (ha) BRN	84.1	0.000	2621	0.000	11,944	58.8	760	1521	1780	0.000	153
Interior waters (ha) RIVERS	236	1027	1106	57.2	1659	117	966	1340	475	32.2	729
Reservoir type	a	a	a	a	a	a	a	a	b	b	b
Sampling periodicity	trianual	trianual	annual	biannual	biannual	annual	trianual	trianual	annual	annual	annual

[a] Riverine reservoirs.

[b] Artificial lake reservoir.

Environmental Variables and Chlorophyll A

From 1996 to 2004, the environmental and biological parameters were measured by the Laboratory of Environment and Applied Chemistry (LABELEC) four times per sampling year, corresponding to spring (April/May), summer (July/August), autumn (October/November) and winter (January/February). The sampling periodicity is indicated in Table 1. All samples were collected at 100 m from the reservoirs's crest, at two different depths: (a) near the surface (approximately 0.5 m depth); and (b) near the bottom (2 m above bottom, only for environmental parameters).

Water temperature, turbidity, conductivity, pH and dissolved oxygen were determined *in situ* using a YSI handheld multiparameter probe (Yellow Spring Instruments). The other environmental variables were determined according to methodologies described by APHA (1995).

To determined soil use dynamics in Douro watershed, i.e. rates of soil use alterations, a geographic information system database was created (ESRI, ArcGIS 9.0), with 13 spatial variables (see Table 1). These use/land cover variables derived primarily from the Corine Land Cover (CLC) from two distinct decades 1990 and 2000 (IGEOE, 2006), additionally the proportions of the predominant CLC classes in the basin (urban areas, intensive and extensive agriculture, natural and semi-natural areas and burned areas) were calculated.

Biological Variables

Like the environmental parameters, phytoplankton samples were collected from 1996 to 2004 at a depth of approximately 0.5 m using a Van Dorn bottle net. Phytoplankton community composition was studied through inverted microscopy, following Utermohl's method (Lund et al., 1958). For the quantification and identification of phytoplankton, samples were fixed in Lugol's solution (1%, v/v) and, when possible, identified to the species level.

Statistical Analysis and Modelling Procedures—STDM

The Stochastic-Dynamic Methodology (StDM) proposed is a sequential modelling process initiated by a multiple regression conventional procedure. However, the fact that the data we considered consisted of n independent variables does not automatically imply that all variables have a significant effect on the magnitude of the dependent variable. Therefore, a stepwise multiple regression analysis (Zar, 1996) was used to test relationships between the soil use dynamics within the watershed (level 1) and the physicochemical variables of the reservoir (level 2) and between these aquatic environmental variables and the phytoplankton metrics (level 3). The developed StDM model includes globally 27 state variables: eleven related to the soil use dynamics (level 1), twelve related to physicochemical variables (level 2) and four related to biological metrics (level 3) (Fig. 2). The conceptual diagram reflects the relationships detected in the multiple regression analysis and on existing relevant regional data sets (Fig. 2, seeCabecinha et al., 2009a, for details). The soil use dataset, the base of the dynamic sub-model of our StDM application (level 1), incorporates real gradients relying on land cover alterations through a decade, from 1990s to 2000s, in the Douro river watershed. This model conceptualisation and validation are fully described in Cabecinha et al. (2009a). In level 3, the dependent variables, selected as representative of the phytoplankton communities, were the number of species of Cyanophyta (blue-green algae), Clorophyta (green algae), Bacillariophyta (diatoms) and Chlorophyll a concentration (Fig. 2). This upper level was influenced by the preceding levels, particularly by environmental variables, including the stochastic ones (Fig. 2).

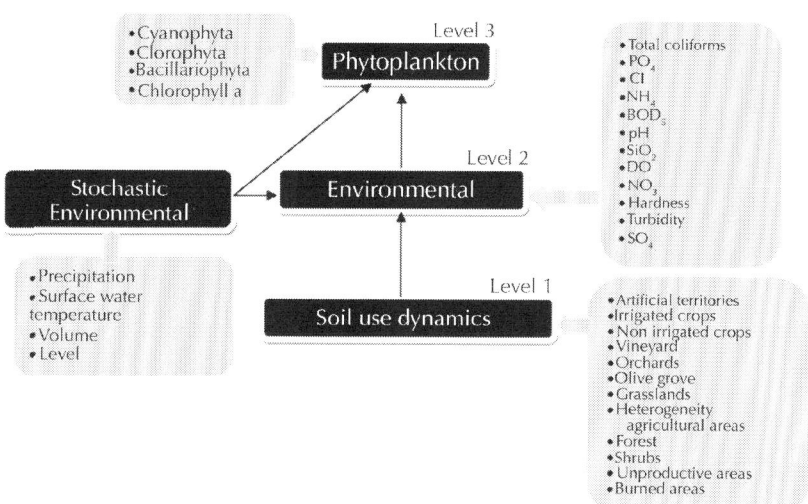

Figure 2: Conceptual diagram of the StDM model used to assess the ecological status of the reservoirs in the Douro watershed, in a multi-scale approach. The model is composed by different sub-models and their interactions: level 1, soil use dynamics model; level 2, predict the responses of the water column environmental variables to changes due to land use changes; level 3, predict the responses of biological variables to changes in environmental variables.

The selected biological metrics largely depend on weather conditions, namely on precipitation and related variables like water residence time (Reynolds, 1984, Basu and Pick, 1996, EPA, 1998 and Dokulil and Teubner, 2000). Therefore, two different complementary equations were calculated for each state variable of phytoplankton groups considered, depending on the monthly precipitation. The categorization in dry or wet months was determined by comparing monthly cumulative precipitations with the reference historical values of monthly precipitation obtained from the period between 1961 and 1990 (Portuguese Weather Institute, 2007). Consequently, the simulation performance of a given state variable results from the calculations of two alternative equations automatically selected in response to the monthly precipitation influence (Cabecinha et al., 2009a).

This multi-level approach gives realism to the interactions considered by incorporating into the model a typical "cascade effect" observed in these holistic processes (Brazner et al., 2007 and Bailey et al., 2007). Therefore, in order to simplify the model structure, only the main key-components were introduced as representative ecological indicators, but which obviously could be complemented by other relevant state variables or other dynamic variables in further applications. The statistical procedure was based on a very complete database, covering true gradients of environmental and biological characteristics of the reservoirs in the Douro basin (Fig. 1), over space and time, the significant partial regression coefficients were assumed as relevant holistic ecological parameters in the dynamic model construction. This is the heart of the philosophy of the StDM. In a holistic perspective, the partial regression coefficients represent the global influence of the environmental variables selected, which are of significant importance on several complex ecological processes. To develop the dynamic model the software STELLA 8.1.4® was used.

Water quality indices and trophic level classification system are useful tools for enhancing communications between scientists, water managers, policymakers and/or the general public. To complement the information about the ecological status of reservoir›s water quality, the trophic state indices based on chlorophyll *a*, total phosphorus and Secchi disk depth outputs were introduced into the model based on Carlson's Trophic Status Index (TSI) (Carlson, 1977). Since nitrogen limitation still classifies a lake along Naumann's nutrient axis, the TSI (TN) (Kratzer et al., 1982) was also introduced in the model (see Table 2for details).

Table 2: A list of possible changes that might be expected in a north temperate lakes or reservoirs as the amount of algae changes along a trophic state gradient (based on Carlson and Simpson, 1996)

TSI	Color	Chl (ug/L)	SD (m)	TP (ug/L)	N (mg/L)	Attributes
<30	Blue	<0.95	>8.0	<6.0	<0.19	Oligotrophy: Clear water, oxygen throughout the year in the hypolimnion

30-40	Green	0.95-2.6	8.0-4.0	6.0-12	0.19-0.36	Oligo-Mesotrophy: Hypolimnia of shallower lakes may become anoxic.
40-50	Yelow	2.6-7.3	4.0-2.0	12-24	0.36-0.74	Mesotrophy: Water moderately clear; increasing probability of hypolimnetic anoxia during summer
50-60	Dark yelow	7.3-20	2.0-1.0	24-48	0.74-1.5	Eutrophy: Anoxic hypolimnia. possible macrophyte problems
60-70	Orange	20-56	0.50-1.0	48-96	1.5-3.0	Eutro-Hypereutrophy: Blue-green algae dominate; problems with algal scums and macrophyte.
>70	Red	56-155	0.25-0.50	96-192	>3.0	Hypereutrophy: Light limited productivity. Dense algae and macrophytes

The model is prepared to produce stochastic simulations based on the monthly stochastic variability of some environmental variables (Random Mode) (see Cabecinha et al., 2009a). Simulations based on stochastic principles take into consideration the random behaviour of some environmental variables with influence on the studied ecological phenomena. The limit values of environmental variables were determined, from the period between December 1995 and December 2004, to discriminate the maximum and minimum values of each stochastic environmental variable, included in the model as a RANDOM function (see Cabecinha et al., 2009a for details). For graphical representations, 10 stochastic simulations were carried out for the simulation period and the average tendencies were calculated for all the state variables considered in this study.

Cellular Automata (CA) and Land use Change

The Cellular Automata (CA) is a dynamic technique that inherently integrates the spatial and temporal dimensions. CA is composed by four basic elements (White and Engelen, 2000 and Scheller and Mladenoff, 2007): (1) *Cell space*: The cell space is composed of individual cell with variable geometric shape, although most CA adopt regular grids to represent such space, which make CA very similar to a raster GIS; (2) *Cell states*: The states of each cell may represent any spatial variable, e.g., several types of land uses; (3) *Time steps*: CA evolves a sequence of discrete time steps and the cells will be updated simultaneously based on transition rules at each step; (4) *Transition rules*: These rules are the heart of a CA that guide its dynamic evolution. A transition rule normally specifies the states of cell before and after updating based on its neighborhood conditions.

Therefore, essential to the simulation of spatial interactions is the representation of spatial information and the neighbourhood structure. In this work square cells of 250 m were used. Spatial interactions among cells were defined by nearest neighbours (see Fig. 3), allowing more flexibility in spatial interactions than vector polygons (Mladenoff and He, 1999).

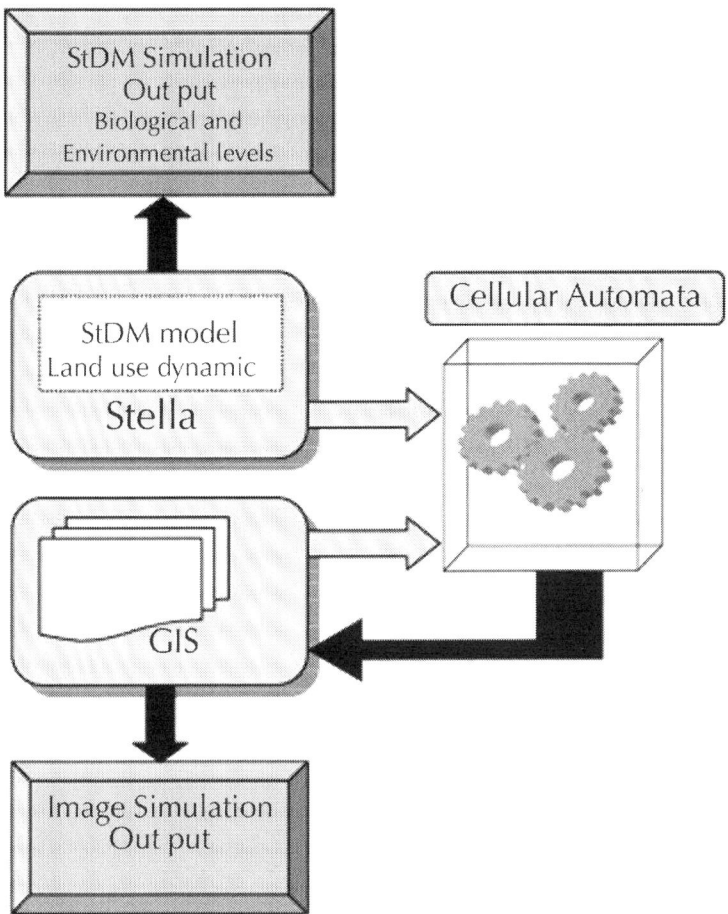

Figure 3: Diagram flow showing the relations between Stella, GIS and the Cellular Automata (CA). These procedures are described in Section 2.5 of the text.

In this work, the transition rules were based in the following criteria: (i) the dynamic land use change sub-model used for our StDM application (the Stella output file) gives the indication for the land use trends; and (ii) a GIS image raster provides the initial values for all land use classes evolved on each simulated dynamic scenario (Fig. 3).

Spatial interactions among cells were defined by nearest neighbours (Fig. 4), allowing more flexibility in spatial interactions than vector polygons (Mladenoff and He, 1999). Therefore, and for demonstration purposes, the simulation algorithm increases, for instance, the area of a given land use (land use A), as indicated by the Stella file, through the identification of the nearest neighbours (land use B), inducing a consequent decrease in the respective area (Fig. 4). In more realistic terms, since an initial type of land use may result in several different land use changes, this study analyzes the various neighborhood of a particular land use change pattern type. Firstly, the algorithm determines the number of elements of each class based in the GIS image raster; secondly, determined the number of elements that is necessary increased in each land use class amplified and the respectively "losing" classes. The process illustrated continued until the objective is complete (Fig. 4).

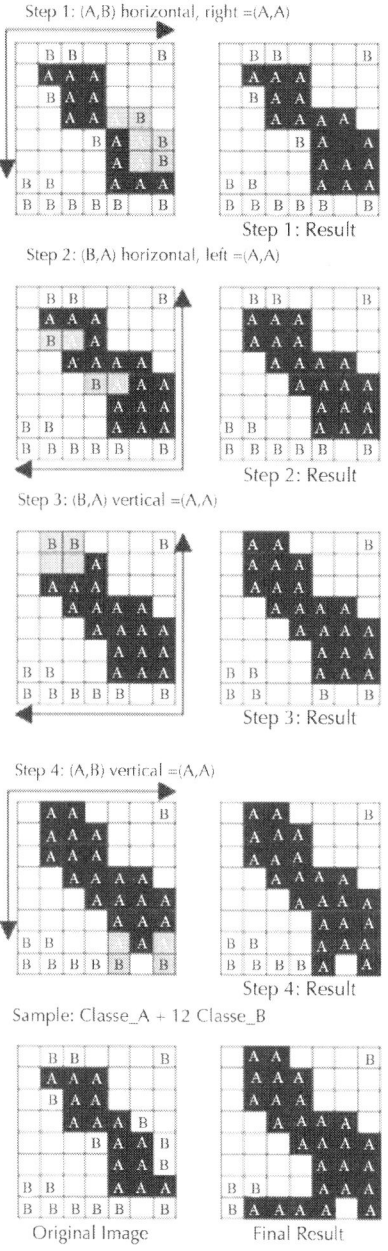

Step 1: (A,B) horizontal, right =(A,A)

Step 1: Result

Step 2: (B,A) horizontal, left =(A,A)

Step 2: Result

Step 3: (B,A) vertical =(A,A)

Step 3: Result

Step 4: (A,B) vertical =(A,A)

Step 4: Result

Sample: Classe_A + 12 Classe_B

Original Image   Final Result

Figure 4: Exemplificative diagram of the application of the Cellular Automata (CA) transitional rules. The simulation algorithm increases the area of a specified land use (land use A), as indicated by the output Stella file, through the identification of the nearest neighbours (land use B, decreased).

Scenario Simulations

The scenario analysis has evolved into a standard methodology in environmental sciences for analyzing the effects of different driving forces and assessing the associated uncertainties. For academic demonstration purposes, a stochastic scenario was developed based on trends of the land cover recorded in the Douro watershed though a decade, from 1990 to 2000 (Corine Land Cover, IGEOE) and on the scenarios identified by the Douro River Basin Management Plan. For this management plan, five different entities were analyzed for possible developments in this region, namely living, working, agriculture, nature and government (INAG, 2008). These different elements were then combined in an integrated spatial scenario. Therefore, for the Torrão watershed the simulated trends in land use change were mainly based in the increasing of natural areas and abandonment of traditional agriculture areas. In Table 3, is described the major characteristics of this scenario.

Table 3: Major characteristics of the scenario adopted for Torrão reservoir taking into account the Douro River Basin Plan (modified from INAG, 2000)

Base simulation scenario
Continuation of current agricultural system with subsidies for farmers, namely for vineyard and olive grove;
Abandonment of traditional agriculture areas, namely in rural regions due to population aging;
Increasing of sustentable agriculture-forest, agriculture-environmental and agriculture-rural investments;
Decreasinsing of the traditional industrial sectors (textile, clothing, shoe, ceramic and metallic products);
Slight increase of residencial areas, mainly associated to urban areas;
Tourism based on the maintenance of the landscape (rural and natural);

> Major increase in natural areas, namely by forestation of burned and unprudutive areas and increasing protected areas.

This scenario was selected to show in detail how the two complementary techniques, StDM and CA, can predicted and capture the expected changes at land use level and how this changes will be reflected in water quality in the Torrão reservoir, namely at physicochemical and phytoplankton levels. In this context, the increase rates used for artificial territory, olive grove and vineyard were calculated according to the ratio: [(soil use area in 2000 − soil use area in 1990)/soil use area in 1990)] (Chaves et al., 2000). The time unit chosen was the month, since it captures in an acceptable way the behaviour of the variables at the lower scales, namely in levels 2 and 3 of our approach. Therefore, the increase rates of the main dynamic land uses, ART TERT, OLV, VIN, BRN, SRB and FRT (see Table 1 for codes explanation), were converted to monthly periods by using the appropriated rate conversions described in Cabecinha et al. (2009a). The simulation period, a decade, was expressed in months, i.e., 120 months. For all state variables, the initial values considered were the data recorded for Torrão's Watershed in December 1995. Since the values of the first month for each period were used as initial values (t_0), the simulations started effectively in January (t_1).

RESULTS AND DISCUSSION

The population growth was assumed to be the most important driver which alters regional economy and hence land use through time. The simulations were performed taking into account the scenario considered, based on realistic trends as stated by the Douro River Basin Management Plan. As a consequence of the populations aging the Douro region has assist an agricultural abandonment in rural areas (INE, 2008). This trend, associated with the diminished of generic agricultural subsidies, is leading to a sharp decrease in the area occupied by traditional agriculture. Nevertheless, the current agricultural system are still funding farmers, namely for vineyard and olive grove. But less profitable agricultural land will be left fallow. Large pieces of agricultural land could be bought up by nature conservation organizations for the development of nature reserves or by private individuals for rural tourism.

Facing this land use change scenario, the estimated responses of pertinent environmental and biological variables from the water column of Torrão reservoir are shown in Fig. 5 and Fig. 6. The simulated trends with alterations at the level of Torrão's Watershed land use have important implications over the aquatic physicochemical parameters and structural composition of the phytoplankton community at the level of the respective reservoir. All the simulations for the environmental and biological variables (Fig. 5 and Fig. 6), as well for the trophic state indices (Fig. 7), reproduce with consistence the seasonal fluctuations expected in this type of reservoirs as a result of such scenario (Wetzel, 2001). Moreover, the scenario of agricultural abandonment in the Torrão catchment has leaded to a decrease in nutrient concentration (NO_3), Cl, hardness and turbidity (Fig. 5). This logic physicochemical response (Jørgensen et al., 2005) had a natural influence on the seasonal pattern of biological variables, with a consequent and credible decrease in Chlorophyll *a* concentration peaks, an expression of potential algal blooms (Heiskanen and Solimini, 2005) (Fig. 5). For the phytoplankton species richness, a dominant resilient pattern was observed, an expectable response since the selected taxonomic groups are composed by species with a high range of sensitivity to environmental stress (Wetzel, 2001). In fact, a relative stable seasonal pattern of phytoplankton succession was observed for all the phytoplankton groups analysed (Fig. 5). These patterns were consistent with the behaviour of these groups facing similar contexts of other temperate lakes and reservoirs (Reynolds, 1984, Mischke, 2003 and Figueiredo et al., 2006). These seasonal successions are strongly conditionated by meteorological and stratification-mixing processes (Wetzel, 2001). Therefore, the variation in the abundance of the total phytoplankton, represented by the chlorophyll*a* content simulations, shown a credible seasonal pattern with an earlier maximum occurring through spring and early summer to a maximum in July (see Fig. 5 and Fig. 6). This pattern is probably typical of many "mesotrophic" temperate lakes and reservoirs (Reynolds, 1984, Wetzel, 2001 and Mischke, 2003). It is supposed that the peaks reflect the coincidence of physically suitable growth conditions, with relatively high concentrations of limiting nutrients shortages of which prevent the attainment of large biomasses during the midsummer period (Reynolds, 1984 and Domingues and

Helena, 2007). Seasonal succession pattern of phytoplankton in temperate reservoirs, such as the case of Torrão, usually involves a winter minimum with species adapted to low light and temperature, a late winter-spring and autumn peaks of diatom richness, followed rapidly by the development of green algae in the spring and finally the transition in late summer and early autumn to a peak of Cyanobacteria (Reynolds, 1984, Wetzel, 2001 and Figueiredo et al., 2006). As shown in Fig. 5 and Fig. 6, the model simulations were able to capture the expected pattern of these variables in this type of reservoirs.

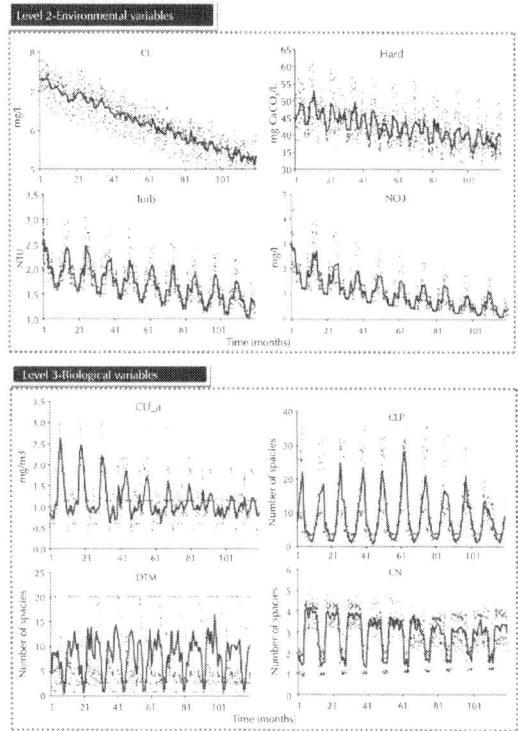

Figure 5: StDM computer simulations for the average values of aquatic environmental (level 2) and biological variables (level 3) from the Torrão reservoir water column through a period of 10 years. The lines result from the average values of 10 monthly stochastic simulations based on the scenario described in Table 3. The specification of the variable codes is expressed in Table 1.

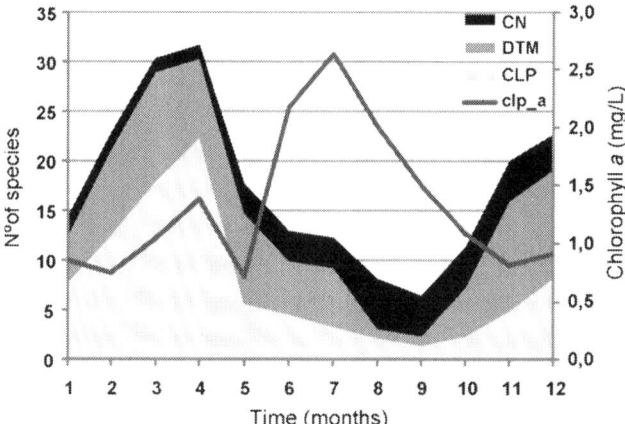

Figure 6: StDM computer simulations showing the seasonal yearly pattern of phytoplankton succession, represented by diatoms (DTM), green algae (CLP) and cyanobacteria (CN) richness and chlorophyll *a* (Clp_a) concentration, in Torrão reservoir. The lines result from the average values of 10 monthly stochastic simulations, for the first year simulated, based on the scenario described in Table 3.

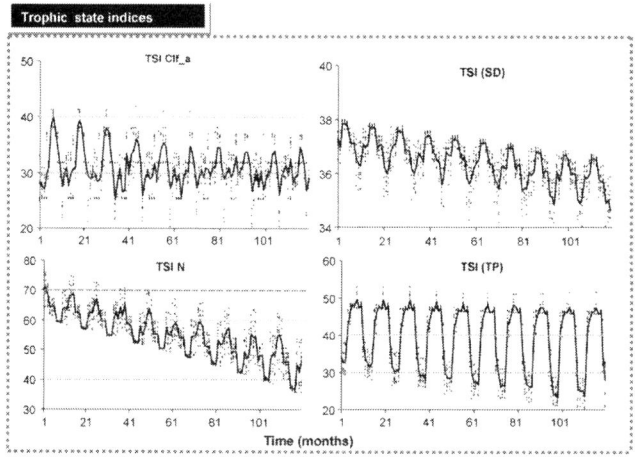

Figure 7: StDM computer simulations for the average values of trophic state indices, namely TSI for chlorophyll *a*, Total phosphorus, Total N and Secchi depth from the Torrão reservoir water column through a period of 10 years. The lines result from the average values of 10 monthly stochastic simulations based on the scenario described in Table 3. The dotted lines represent the limits of the different reservoir trophic state levels, from oligotrophy (<30) to hypereutrophy (>70) (see Table 2 for details).

Relatively to the Trophic State Indices (TSI), they reflected well the meaning of the trends simulated, namely as a result of the decrease of non-point source pollution due to the abandonment of traditional agricultural areas, as well as by forestation of burned and unproductive areas (Jørgensen et al., 2005). As a consequence, the TSI simulated for Chlorophyll a, Total N and Secchi depth, show a consistent decrease throughout the simulation period, mostly reflecting trophic state level changes (Fig. 7). With regard to the TSI (N) the reservoir trophic level changes from hypereutrophic to an oligo-mesotrophic state, which is a realistic response to such scenario, namely by decreasing the anthropogenic impacts associated to agricultural practices. The TSI for total phosphorus concentrations exhibits only a slightly decrease, may be because the phosphorus retention capacity of deep reservoirs is very high in temperate reservoirs (Wetzel, 2001 and Jørgensen et al., 2005). This capacity could reach 80% of the upflow (Jørgensen et al., 2005).

As this StDM model have been published and statistically validated in a previous work (Cabecinha et al., 2009a) it was deemed robust and suitable for our purpose. Overall, the performance of the StDM simulations shows realism in capturing the behavioural patterns for the generality of the relevant metrics adopted. Therefore, the state variables used in the construction of the StDM model reflect well the shift of the environmental characteristics towards known holistic conditions and are capable of responding with credibility to the dynamics of the underlying ecological "cascade" processes (that are implicit in a multi-scale perspective). These results showed that the aquatic metrics selected, as state variables, were not indifferent to changes in the ecological conditions, namely when conditions were changed by man-induced disturbances at the watershed level. The relevant ecological drifts simulated are in agreement with real observations and other studies that investigated the biological consequences of similar environmental changes (Wetzel, 2001, Jørgensen et al., 2005 and Cabecinha et al., 2009b).

Several studies have found the landscape structure to be the main factor influencing water quality in streams and reservoirs (Allan et al., 1997, Cifaldi et al., 2004 and Soininen, 2007). This influence of land use is scale dependent and varies in time and space (Buck et al., 2004). This has been shown at the global scale, as well as at the regional and local scales for catchments dominated by agricultural land use, for forested areas and for heterogeneous multifunctional landscapes (Chen et al., 2001 and Buck et al., 2004). In designing landscapes for reservoir management the great challenge is to make the knowledge gleaned about ecological processes and species response accessible to the planning process. Often this information is expressed mathematically, which is not meaningful or accessible to landscape planners. Therefore, in order to make this knowledge available to the planning process, the scenario adopted in this study must be expressed spatially in a format suitable for scenario-testing. With this objective, the same dynamic model outputs, used to obtain the StDM simulations, enable us to simulate, by using the CA interface with GIS, the spatial changes described for the Torrão watershed (Fig. 8). As shown in Fig. 8, when simulating land uses trends, increasing one type (e.g., Olive grove, vineyard or forest) within the studied region will inevitably decrease the others, such as the traditional agricultural, burned and unproductive areas. Although the current CA approach is rather simple, the merit lies in its ability in revealing the spatial dynamic of land use evolution established by dynamic outputs.

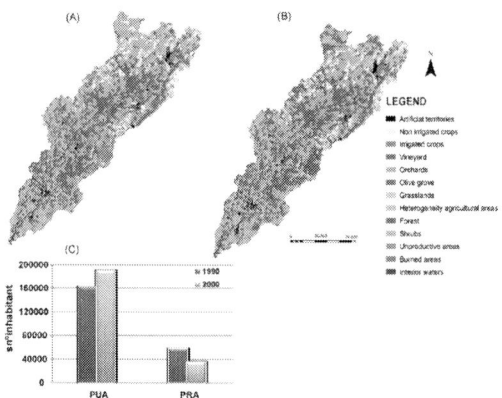

Figure 8: Cellular Automata simulation, through a period of 10 years, of land use change in Torrão watershed, based on the scenario described in Table 3. (A) represents the initial state, (B) the final simulation output, 10 years later and (C) represents the concomitant human population tendency (no. of inhabitants), from 1990 to 2000, in predominant urban (PUA) and rural (PRA) areas from the Torrão's watershed.

CONCLUSIONS

Nowadays, in monitoring and management programs, the construction of predictive tools for ecological management, namely in terms of cost and speed of reliable assessment results, is crucial. In this scope, we believe that our present proposal will provide the development of a true management tool, namely taking into account stochastic/random phenomena that characterize the real ecological processes (Van der Meer et al., 1996). Moreover, the methodology presented offers a unifying approach with the spatial background and the multi-level connections that gives realism to the interactions considered by incorporating a typical "cascade effect" observed in the dynamic of the studied reservoirs. On the other hand, since the watersheds are characterized by a high degree of heterogeneity in space and time, influenced by many interacting factors and by feedback mechanisms, this multi-scale approach is particularly helpful to capture these multi-factor influences in relevant planning scenarios.

When compared to other modelling methodologies, such as Artificial Intelligence (Džeroski et al., 1997 and Kuo et al., 2006), our methodology is more intuitive, namely in mathematical terms, providing easy explanations for the underlying relations between independent and dependent variables and because is based on conventional linear methods that allowed a more direct development of testable hypotheses.Džeroski et al. (1997) referred that models produced in the form of rules, based on machine learning approaches, are transparent and can be easily understood by experts. The integrative application of the two complementary techniques, StDM and CA, exhibits these structural qualities but provides also simple, suitable and intuitive outputs, easily interpreted by non-experts (ranging from resource users to senior policy makers). The selection of new scenarios in both techniques is quick and relatively easy to the end-user and allows not only the direct expression of spatial changes to be visualised, but also for the consequences of the changes for water quality dynamics to be assessed.

Therefore, this study seems to represent a useful contribution for the holistic implementation of the WFD, namely for integrated assessments of the reservoirs ecological status within the environmental gradients or "data space" monitored. Since the ultimate goal was to produce simulation models that permit the creation of spatially explicit ecological patterns we believe that the present approach provide a useful starting point to develop more global techniques in the scope of this research area. In fact, the methodology presented in this paper is an example of how this can be achieved, by creating expeditious interfaces with GIS, which make the modelling outputs more instructive and intuitive to decision-makers and environmental managers (Costanza and Voinov, 2003).

ACKNOWLEDGEMENTS

The authors are indebted to all the colleagues from the GAPI of the University of Trás-os-Montes e Alto Douro (UTAD), namely to Dr. Jorge Machado who assisted in the StDM model patent process (PT103753, pat. pend). We also thank to Dr. D. Fernandes Barbosa and Eng Pedro Silva-Santos for their helpful contributions on early GIS and stochastic-dynamic outputs, respectively. Additionally, we would like to thank the LABELEC staff for the environmental and phytoplankton data, namely to Eng. Lourenço Gil.

REFERENCES

1. Alberti, M., Booth, D., Hill, K., Coburn, B., Avolio, C., Coe, S., Spirandelli, D., 2007. The impact of urban patterns on aquatic ecosystems: an empirical analysis in Puget lowland sub-basins. Landscape and Urban Planning 80 (4), 345–361.

2. Allan, J.D., Erickson, D.L., Fay, J., 1997. The influence of catchment land use on stream integrity across multiple spatial scales. Freshwater Biology 37, 149–161.

3. APHA, 1995. Standard Methods for the Examination of Water and Wastewater, 19th ed. American Public Health Association, Washington, DC.

4. Bailey, R.C., Reynoldson, T.B., Yates, A.G., Bailey, J., Linke, S., 2007. Integrating stream bioassessment and landscape ecology as a tool for land use planning. Freshwater Biology 52, 908–917.

5. Basu, B.K., Pick, F.R., 1996. Factors regulating phytoplankton and zooplankton biomass in temperate rivers. Limnology and

Oceanography 41 (7), 1572–1577.

6.   Brazner, J.C., Danz, N.P., Niemi, G.J., Regal, R.R., Trebitz, A.S., Howe, R.W., Hanowski, J.M., Johnson, L.B., Ciborowski, J.J.H., Johnston, C.A., Reavie, E.D., Brady, V.J., Sgro, G.V., 2007. Evaluation of geographic, geomorphic and human influences on Great Lakes wetland indicators: a multi-assemblage approach. Ecological Indicators 7, 610–635.

7.   Buck, O., Niyogi, D.K., Townsend, C.R., 2004. Scale-dependence of land use effects on water quality of streams in agricultural catchments. Environmetal Pollution 130 (2), 287–299.

8.   Cabecinha, E., Cortes, R., Cabral, J.A., 2004. Performance of a Stochastic-Dynamic Modelling methodology for running waters ecological assessment. Ecological Modelling 175 (3), 303–317.

9.   Cabecinha, E., Silva-Santos, P., Cortes, R., Cabral, J.A., 2007. Applying a stochasticdynamic methodology (StDM) to facilitate ecological monitoring of running waters, using selected trophic and taxonomic metrics as state variables. Ecological Modelling 207, 109–127.

10. Cabecinha, E., Cortes, R., Pardal, M.A., Cabral, J.A., 2009a. A stochastic dynamic methodology (StDM) for reservoir water quality management: the validation of a multi-scale approach in a South Europe basin (Douro, Portugal). Ecological Indicators 9 (2), 240–255, doi:10.1016/j.ecolind.2008.05.010.

11. Cabecinha, E., Cortes, R., Cabral, J.A., Ferreira, T., Lourenc̜o, M., Pardal, M.A., 2009b. Multi-scale approach using phytoplankton as a first step towards the definition of reservoirs ecological status. Ecological Indicators 9 (2), 329–345, doi:10.1016/j.ecolind.2008.04.006.

12. Cabral, J.A., Rocha, A., Santos, M., Crespí, A.L., 2007. A stochastic dynamic methodology (SDM) to facilitate handling simple passerine indicators in the scope of the agri-environmental measures problematics. Ecological Indicators 7, 34– 47.

13. Carlson, R.E., 1977. A trophic state index for lakes. Limnology and Oceanography 22, 361–369.

14. Carlson, R.E., Simpson, J., 1996. A Coordinator's Guide to Volunteer Lake Monitoring Methods. North American Lake Management Society, p. 96.

15. Chaloupka, M., 2002. Stochastic simulation modelling of southern Great Barrier Reef green turtle population dynamics. Ecological Modelling 148, 79–109.

16. Chaves, C., Maciel, E., Guimarães, P., Ribeiro, 2000. Instrumentos estatísticos de apoio à economia: conceitos básicos. McGram-Hill, Lisboa, Portugal.

17. Chen, L., Wang, J., Bojie, F., Qiu, Y., 2001. Land-use change in a small catchment of northern Loess Plateau China. Agriculture Ecosystems and Environment 86, 163–172.

18. Cifaldi, R.L., Allan, J.D., Duh, J.D., Brown, D.G., 2004. Spatial patterns in land cover of exurbanizing watersheds in southeastern Michigan. Landscape and Urban Planning 66, 107–123.

19. Costanza, R., Voinov, A., 2003. Introduction: spatially explicit landscape simulation models. In: Costanza, R., Voinov, A. (Eds.), Landscape Simulation Modeling, A Spatially Explicit Dynamic Approach. Springer Verlag, New York, pp. 3–20.

20. Dokulil, M.T., Teubner, K., 2000. Cyanobacterial dominance in lakes. Hydrobiologia 438, 1–12.

21. Domingues, R.B., Helena, Galvão, 2007. Phytoplancton and environmental variability in a dam regulated temperate estuary. Hydrobiologia 586, 117–134.

22. Dzeroski, S., Grbovic, J., Walley, W.J., Kompare, B., 1997. Using machine learning ˇ techniques in the construction of models. 2, data analysis with rule induction. Ecological Modelling 95 (1), 95–111.

23. EPA, 1998. Lake and Reservoir Bioassessment and Biocretirea U.S. Environment Protection Agency. Technical Guidance document. Oficce of water, Washington DC, EPA/841-B-98-007.

24. European Union, 2000. Directive 2000/60/EC of the European Parliament and of the Council Establishing a Framework for the Community Action in the Field of Water Policy. European Commission, off. Journal of European Community L327, 1.

25. Figueiredo, D.R., Reboleira, A.S., Antunes, S.C., Abrantes, N., Azeiteiro, U.M., Gonc̨ alves, F., Pereira, M.J., 2006. The effect of environmental parameters and cyanobacterial blooms on phytoplankton dynamics of a Portuguese temperate lake. Hydrobiologia, doi:10.1007/s10750-006-0196-y.

26. Firebaugh, M.W., 1988. Artificial Intelligence A Knowledge-Based Approach. Boyd and Fraser Publishing Company, Boston, p. 740.

27. Heiskanen, A., Solimini, A.G., 2005. Relationships between pressures, chemical status, and biological quality elements. Analysis of the current knowledge gaps for the implementation of the Water Framework Directive. Joint Research Centre, European Commission.

28. IGEOE, Instituto Geográfico do Exército (Geografic Military Institute), 2006. Corine Land Cover 1990 and 2000. http://www.igeoe.pt/.

29. INAG, 2008. Portuguese Water National Institute. Douro river Catchment Plan. (http://www.inag.pt/inag2004/port/a intervencao/planeamento/pbh/pbh02. html.

30. INE, 2008. Portuguese Statistic National Institute. http://www.ine.pt/portal/ page/portal/PORTAL INE/DEstatisticos.

31. Jørgensen, S.E., 1994. Models as instruments for combination of ecological theory and environmental practice. Ecological Modelling 75/76, 5–20.

32. Jørgensen, S.E., 1995. State of the art of ecological modelling in limnology. Ecological Modelling 78, 101–115.

33. Jørgensen, S.E., 1999. State-of-the-art of ecological modelling with emphasis on development of structural dynamic models. Ecological Modelling 120, 75–96.

34. Jørgensen, S.E. (Ed.), 2001. Fundamentals of Ecological Modelling, 3rd ed. Elsevier, Amsterdam.

35. Jørgensen, S.E., 2005. Ecological Modelling: editorial overview 2000–2005. Ecological Modelling 188 (2–4), 137–144.

36. Jørgensen, S.E., Löffler, H., Rast, W., Straskraba, M., 2005. Lake and Reservoir Man- ˇ agement. Developments in Water Science. Elsevier, p. 502.

37. Klaver, G., van Os, B., Negrel, P., Petelet-Giraud, E., 2007. Influence of hydropower dams on the composition of the suspended and riverbank sediments in the Danube. Environmental Pollution 148 (3), 718–728.

38. Kratzer, C.R., Brezonik, P.L., Osgood, R.A., 1982. A Carlson-type trophic state index for nitrogen in Florida Lakes" by Charles R. Kratzer and Patrick L. Brezonik. Journal of American Water Resources Association 18 (2), 343–344.

39. Kuo, J.-T., Wang, Y.-Y., Lung, W.-S., 2006. A hybrid neural-genetic algorithm for reservoir water quality management. Water Research 40 (7), 1367–1376.

40. Lund, J.W.G., Kipling, C., Le Cren, E.D., 1958. The invertited microscope methods of estimating algal numbers and the statistical basis of estimation by counting. Hydrobiologia 11, 143–170.

41. Malafant, K.W.J., Fordham, D.P., 1998. In: Uso, J.L., Brebbia, C.A., Power, H. (Eds.), GIS DSS and Integrated Scenario Modelling Frameworks for Exploring Alternative Futures. Ecosystems and Sustainable Development. Computational Mechanics Publications, Southampton, England, pp. 669–678.

42. Mischke, U., 2003. Cyanobacteria associations in shallow polytrophic lakes: influence of environmental factors. Acta Oecologica 24, S11–S23.

43. Mladenoff, D.J., He, H.S., 1999. Design, behavior and application of LANDIS, an objectoriented model of forest landscape disturbance and succession. In: Mladenoff, D.J., Baker, W.L. (Eds.), Spatial Modeling of Forest Landscape Change. Cambridge University Press, Cambridge, UK, pp. 125–162.

44. Portuguese Weather Institute, 2007. http://web.meteo.pt/pt/clima/clima.jsp. Reynolds, C.S., 1984. The Ecology of Freshwater Phytoplankton. Series. Cambridge Studies in Ecology.

45. Santos, M., Cabral, J.A., 2004. Development of a stochastic dynamic model for ecological indicators' prediction in changed Mediterranean agroecosystems of north-eastern Portugal. Ecological Indicators 3, 285–303.

46. Santos, M., Vaz, C., Travassos, P., Cabral, J.A., 2007. Simulating the impact of socioeconomic trends on threatened Iberian wolf

populations (Canis lupus signatus) in North-eastern Portugal. Ecological Indicators 7, 649–664.

47. Schauser, I., Lewandowski, J., Hupfer, M., 2003. Decision support for the selection of an appropriate in-lake measure to influence the phosphorus retention in sediments. Water Research 37 (4), 801–812.

48. Scheller, R., Mladenoff, D., 2007. An ecological classification of forest landscape simulation models: tools and strategies for understanding broad-scale forested ecosystems. Landscape Ecology, doi:10.1007/s10980-006-9048-4.

49. Silva-Santos, P.M., Pardal, M.A., Lopes, R.J., Múrias, T., Cabral, J.A., 2006. A stochastic dynamic methodology (STDM) to the modelling of trophic interactions, with a focus on estuarine eutrophication scenarios. Ecological Indicators 6, 394–408.

50. Silva-Santos, P., Pardal, M.A., Lopes, R.J., Múrias, T., e Cabral, J.A., 2008. Testing the Stochastic Dynamic Methodology (StDM) as a management tool in a shallow temperate estuary of south Europe (Mondego Portugal). Ecological Modelling 210:, 377–402.

51. Soininen, J., 2007. Environmental and spatial control of freshwater diatoms—a review. Diatom Research 22, 473–490.

52. Van der Meer, J., Duin, R.N.M., Meininger, P.L., 1996. Statistical analysis of long-term monthly Oystercatcher Haematopus ostralegus counts. Ardea 84A, 39–55.

53. Wetzel RG, 2001. Limnology—Lake and River Ecosystems. Academic Press, San Diego, p. 1006.

54. White, R., Engelen, G., 2000. High-resolution integrated modeling of the spatial dynamics of urban and regional systems, Computer. Environment and Urban System 24, 383–400.

55. Yang, Q., Li, X., Shi, X., 2008. Cellular automata for simulating land use changes based on support vector machines. Computers & Geosciences 34, 592–602.

56. Zar, J.H. (Ed.), 1996. Biostatistical Analysis, 3rd ed. Prentice-Hall International Inc., Englewood Cliffs, NJ.

Longitudinal Hydrodynamic Characteristics in Reservoir Tributary Embayments and Effects on Algal Blooms

Huichao Dai[1], Jingqiao Mao[1], Dingguo Jiang[2],
and Lingling Wang[1]

[1]State Key Laboratory of Hydrology-Water Resources and Hydraulic Engineering, Hohai University, Nanjing, China,

[2]College of Civil and Hydroelectric Engineering, China Three Gorges University, Yichang, China

ABSTRACT

Three Gorges Reservoir (TGR) is one of the largest man-made lakes in the world. Since the impoundment in 2003, however, algal blooms have been often observed in the tributary embayments. To control the algal blooms, a thorough understanding of the hydrodynamics (e.g., flow regime, velocity gradient, and velocity magnitude and direction) in the tributary embayments is particularly important. Using a calibrated three-dimensional hydrodynamic model, we carried out a hydrodynamic analysis of a typical tributary embayment (i.e., Xiangxi

Bay) with emphasis on the longitudinal patterns. The results show distinct longitudinal gradients of hydrodynamics in the study area, which can be generally characterized as four zones: riverine, intermediate, lacustrine, and mainstream influenced zones. Compared with the typical longitudinal zonation for a pure reservoir, there is an additional mainstream influenced zone near the mouth due to the strong effects of TGR mainstream. The blooms are prone to occur in the intermediate and lacustrine zones; however, the hydrodynamic conditions of riverine and mainstream influence zones are not propitious for the formation of algal blooms. This finding helps to diagnose the sensitive areas for algal bloom occurrence.

INTRODUCTION

Three Gorges Reservoir (TGR) is a typical huge man-made lake, located in the upper Yangtze River, China. It is also one of the largest impounded reservoir systems in the world, with a normal pool level of 175 m and a total reservoir storage capacity of 39.3 billion m³. TGR currently faces the dual challenge of successfully performing the necessary tasks (flood control, hydropower generation, and navigation) while at the same time minimizing the negative environmental impacts [1]. Although the reservoir mainstream currently maintains the mesotrophic level, algal bloom events occur episodically in the tributary embayments [2]. For example, there were 6 bloom events observed at some embayments in 2004, 19 in 2005, and 10 from February to March 2006, respectively [3]. Among them is the representative Xiangxi Bay located not far upstream to the TGR (Fig. 1).

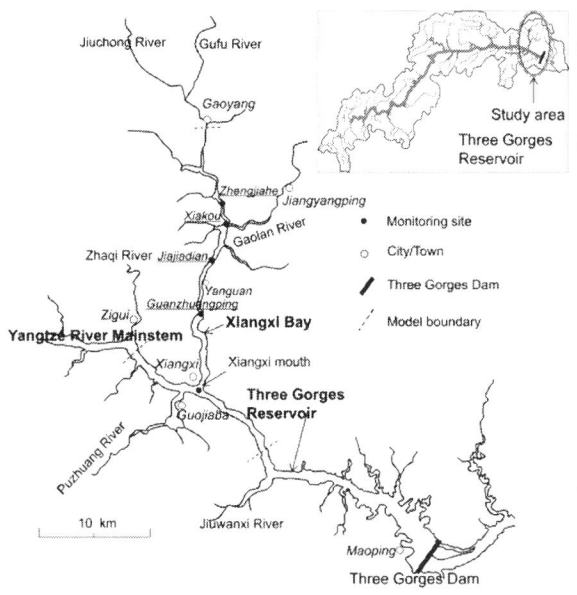

Figure 1: Map of Xiangxi Bay of Three Gorges Reservoir and location of model boundaries.

Algal bloom management must be based on a thorough understanding of the hydrodynamics, because it is commonly believed that (*i*) blooms mainly occur in eutrophic waterbodies under favorable hydrodynamic and meteorological conditions [4], [5], and (*ii*) algal bloom mitigation is determined by pollution loading and hydrodynamic conditions [6]. Spatial pattern analysis is a useful tool for reservoir ecosystem management [7]–[9]. Flow regimes of reservoirs are usually more complex than that of natural lakes or rivers, and are periodically affected by reservoir operations. Nevertheless, for a reservoir mainstream that has no large lateral tributaries, it is found that there are longitudinal gradients in the physical, chemical, and biological properties. Such a reservoir can be generally characterized as three longitudinal zones (Fig. 2a): upper riverine zone, middle transitional zone and lower lacustrine zone [10], [11]. The riverine zone is defined at the upper reach that has a narrow and channelized basin with relatively higher flow rates; the transitional zone is located at the middle reach that has a relatively broader and deeper basin with reduced flow rates; the lacustrine zone refers to the lower part immediately upstream to the dam, which has a broader, deeper and lake-like basin with lower flow rates. For example, according to the technical guideline proposed by the Ministry of Environmental Protection of China [12], the longitudinal zonation of TGR mainstream may be simply determined based on the monthly mean flow velocities: U>0.03 m/s, riverine zone; 0.03 m/s>U>0.01 m/s, intermediate zone; U<0.01 m/s, lacustrine zone.

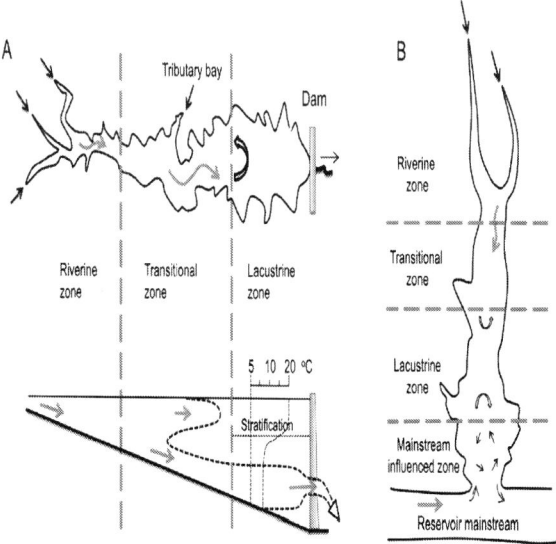

Figure 2: Schematic illustration of longitudimal zonation in a reservoir (the left modified from Ref [10]), and in a tributary embayment (right).

Few studies have discussed the differences of longitudinal zonation between tributary embayments and normal reservoir mainstreams, from the point of view of hydrodynamic analysis. Over the last decade, some hydro-environment research has been carried out in the representative embayment of TGR, Xiangxi Bay. Most studies focused on the field investigation of the hydrological, hydrodynamic, and ecological conditions [13]–[15]. Some literatures studied the longitudinal patterns of phytoplankton and macroinvertebrate community [14], [16]–[17]. Whereas the general eutrophication process has been modeled [18], however, the cause-and-effect relationship between longitudinal hydrodynamic characteristics and algal bloom events are not well understood.

Numerical modeling is a powerful tool for quantitatively analyzing hydrodynamic characteristics, because it can be utilized to assess a range of alternatives (e.g., reservoir operations, discharge management, and bathymetry changes) after proper calibration and validation. It covers a wide range of models from the simple 1D Saint-Venant equations to some complex 3D turbulence models. Among the widely used and recommended hydrodynamic models are some general, flexible and dynamically-coupled frameworks [19]–[24]. Due to their flexible grid systems (e.g., curvilinear or unstructured grids), these models can be applied to different types of waterbodies in one, two, and three dimensions. In addition, for a relatively long and narrow waterbody, a 2D longitudinal-vertical model [25] is also a good choice. The models above-mentioned usually solve the vertically hydrostatic, free surface, turbulent averaged equations of water motions. Although the robustness of these models has been verified by a number of case studies, the model application still requires a good knowledge of computational fluid dynamics as well as comprehensive observational data.

The study is the first step towards developing effective algal bloom management strategies for TGR. We aim to examine the longitudinal hydrodynamic characteristics of reservoir tributary embayments, and analyze the effects on the algal blooms. Using the case study of Xiangxi Bay, a three-dimensional (3D) hydrodynamic simulation is carried out based on the general Delft3D framework. Based on the hydrodynamic analysis, a simple longitudinal zonation method is designed for reservoir tributary embayments. Through a comparative study between the hydrodynamic results and available algal bloom data, the relation between longitudinal zonation and algal bloom events is discussed.

MATERIALS AND METHODS

Ethics Statement

No specific permits were required for the described field studies. The location studied is not privately-owned or protected in any way, and does not include a national park or other protected area of land. The field studies did not involve endangered or protected species.

Study Area

Xiangxi River is one of the longest tributaries of TGR, located 34.5 km upstream from the dam (Fig. 1). The drainage area is around 3099 km^2 (110°25′–111°06′E, 30°57′–31°34′N). According to the statistical data, the average annual air temperature of the study area is 16.6°C, and the average annual rainfall is 1016 mm [26]. It has three main subtributaries, and the average inflow discharge is 40.18 m^3/s [26]. After the impoundment in June 2003, a river-like embayment was formed at Xiangxi River. Consequently, the flow movements in the eutrophic bay are generally weak (around 1–10 mm/s normally, and 0.1 m/s of flood peak), possibly leading to nuisance algal blooms [13], [15].

Choice of Hydrodynamic Model

The general model framework, Delft3D, is applied and configured to simulate the flow regimes in Xiangxi Bay. We selected Delft3D as the platform because: (*i*) it is a widely accepted modeling tool for hydrodynamic simulations in rivers, lakes, reservoirs and coastal waters[27]–[28], and (*ii*) it can model the waters using curvilinear structured grids and includes the ability to simulate drying and wetting of shallow waters, which is especially suitable for river-like embayments, and (*iii*) it has been carefully validated against a series of benchmark experiments by the authors and collaborators, such as long wave propagation [29], salinity intrusion [30], stratification [31] and flushing time of waterbodies [32].

The governing equations are the unsteady 3D shallow-water equations derived from the Reynolds-averaged Navier-Stokes equations for turbulent flows, consisting of the continuity, momentum, and the hydrostatic equations. The vertical eddy diffusivity coefficient is computed following the standard two-equation k-ε turbulence model [33]. Thermal effects may play an important role in the flow regime and water quality distribution. Here, water temperature is computed by solving the transport equation which includes the net heat flux across the surface [20]. The above-mentioned governing equations are numerically solved by a combination of central and upwind spatial discretization (finite difference techniques). Specifically, the staggered Arakawa C-grid is used, where the velocity components are perpendicular to the cell faces, while the water level (pressure) is specified at the cell center. When solving the discretized equations in time, a two-step alternating direction implicit scheme is applied. More details of the numerical solution can be found in Delft Hydraulics [20].

Data Collection

Suitable geometry, bathymetry, initial and boundary conditions are required for a river-like reservoir simulation. The geometry and bathymetry data provided by the Technology and Environmental Protection Department of China Three Gorges Corporation (CTG-TEPD) are used to supply basic information for modeling. Discharge and water level data also obtained from CTG-TEPD are used to provide the necessary initial and boundary conditions. To calibrate and validate the model, field data (i.e., water level, flow velocity, and water temperature) at the stations of Zhengjiahe, Xiakou and Xiangxi mouth (Fig. 1) during the study period are collected. The meteorological data used for water temperature computation are obtained from China Meteorological Data Sharing Service System (see http://cdc.cma.gov.cn/). In addition, to investigate the relation between hydrodynamic regimes and algal abundance of the bay, we conducted the water quality surveys in the study area, approximately on a biweekly basis at 4 sites in Spring 2005 (Fig. 1). With regard to the data quality, high frequency field observations of water levels, discharges and hydrometeorological parameters (sampling interval $\Delta t = 1$ h, 12 h, 6 h, respectively) are collected for boundary and initial conditions. However, relatively limited field data within the study area ($\Delta t =$ two weeks or more) are available for calibration and validation.

Model Configuration of Xiangxi Bay

The computational domain for Xiangxi Bay ranges from the upstream Gaoyang Town to the Xiangxi mouth which is defined at the junction between the tributary and the reservoir mainstream (~32 km long; see Fig. 1). Considering the possible effects of the reservoir mainstream on the bay, a section of the Yangtze River is also modeled (~18 km long; dash lines in Fig. 1). The boundary-fitted curvilinear coordinate and sigma-coordinate are adopted to fit the natural boundaries of the study area. Extensive numerical experiments were carried out using different grid resolutions, ordered from coarse to fine, to test the sensitivity of model results to horizontal model grid. The grid system eventually used includes 867×28 grid nodes for Xiangxi Bay and 359×13 grid nodes for the mainstream part. The grid cell sizes vary from 9.8 m to 58 m in the bay, and 20–70 m in the mainstream. The grid in the Xiangxi Bay has been specially refined to study the longitudinal hydrodynamic characteristics. In the vertical direction, 10 uniformly distributed sigma-layers are employed. A sensitivity test was conducted to determine the proper layers using 5, 10 and 15 sigma-layers, respectively; it is found that when modeling the thermal distribution, the cases of 10 layers can provide more detailed information and retain reasonable computational economy.

At the upstream boundaries of Xiangxi Bay and the reservoir mainstream, the measured daily discharge data are specified, and the observed water level records immediately upstream of the Three Gorges Dam are considered to be the downstream boundary conditions (e.g., during Spring 2005; Fig. 3). A free slip condition is applied along the solid boundaries, and the transformed vertical velocity at the free surface and bottom are set to be zero. In addition, a map of spatially varying Manning roughness coefficients is calibrated: 0.03 for the mainstream, 0.026 for the bay mouth, and 0.02 for other areas. Meteorological data used for thermal computation include air temperature, radiation, pressure, humidity, precipitation and wind (speed and direction); here, the continuous air temperature and light data are provided (Fig. 4), while the mean values of other parameters are specified using the data observed at a station ~40 km downstream from the embayment.

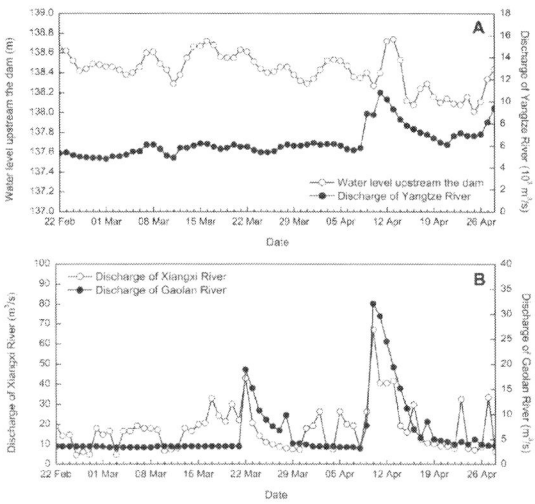

Figure 3: Hydrodynamic boundary conditions applied to the model: (a) observed discharges of Yangtze River and water levels immediately upstream of the dam, and (b) discharges of Xiangxi River and Gaolan River.

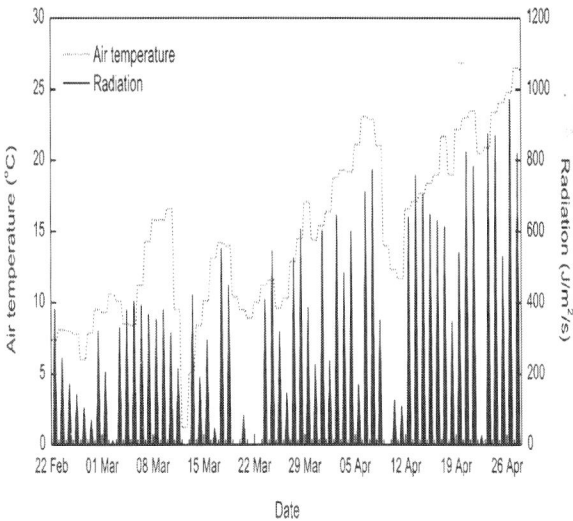

Figure 4: Observed air temperature and solar radiation at Xiangxi Bay during 22 Feb–28 Apr 2005.

Considering the Three Gorges Reservoir operation is subject to the comprehensive requirements of flood control, power generation, and navigation, the water levels can vary widely during a year and may fluctuate substantially within a short time (e.g., ~0.5 m in mid-April 2005). Therefore, during the computational process, the wetting and drying technique is used for describing the moving shores: the process of drying and flooding is represented by removing grid points from the flow domain when the water level locally falls below a certain threshold (drying cells), and by again adding grid points into the flow domain when the local water level rises above a second threshold (wetting cells). A 30-day spin-up period from "cold start" is used before the actual hydrodynamic simulation starts, for which a zero flow velocity field is assumed first while measured data are adopted at the boundaries. To reduce the spin-up time, however, the initial temperature distributions for the spin-up period are based on the spatially interpolated data, i.e., using "warm start" instead of "cold start".

RESULTS

Model Verification

To ensure the model can properly simulate the hydrodynamic characteristics of the tributary embayment, a calibration and validation procedure is performed: first, the model framework-Delft3D is fully tested against a series of benchmark experiments [29]–[32]; second, the observation of general flow pattern (in particular for the shear flow phenomenon) is then adopted to qualitatively evaluate the model performance; lastly, the model is quantitatively calibrated and verified with field data of hydrodynamics (direct validation) and water quality (indirect validation) observed in different locations and time periods.

Two representative cases are demonstrated herein. For the step of quantitative calibration, the observed hydrodynamic process associated with the blooms during February-May 2007 is used. The observed data during February-April 2005 are then adopted to validate it again. Although the hydrodynamic conditions changes significantly from case to case, both cases show reasonable agreement of the model results and the observed data. The calibration performance is judged by comparison of the computed and measured flow velocities at a site within Xiangxi Bay (Station Zhengjiahe; Fig. 1) during Spring 2007. Fig. 5(a) and (b) show good agreement of the model with the measured water level and speed data. The mean absolute relative error

$$\left(MARE = \frac{1}{n}\sum_{i=1}^{n}\left(\left|Observed - Modelled\right| / Observed \right) \right)$$

between the predicted and measured water levels is 0.024%. The MARE of depth-averaged velocity magnitude at the same site is 5.43%. Water temperature data collected at Station Xiakou (in Xiangxi Bay; Fig. 1) during Spring 2007 are used to evaluate the thermal profile results. It is shown that the model could accurately capture the onset of stratification and surface-bottom temperature differences (MARE = 1.74%; Fig. 6a). Using the calibrated coefficients while changing the boundary conditions, the computed currents in the TGR mainstream are verified with the field data measured at a station near the mouth (Station Xiangxi mouth; Fig. 1) during Spring 2005. Fig. 5(c) and (d) show good agreement of the model with the measured water level and speed data (MARE = 0.034% and 3.29%, respectively).

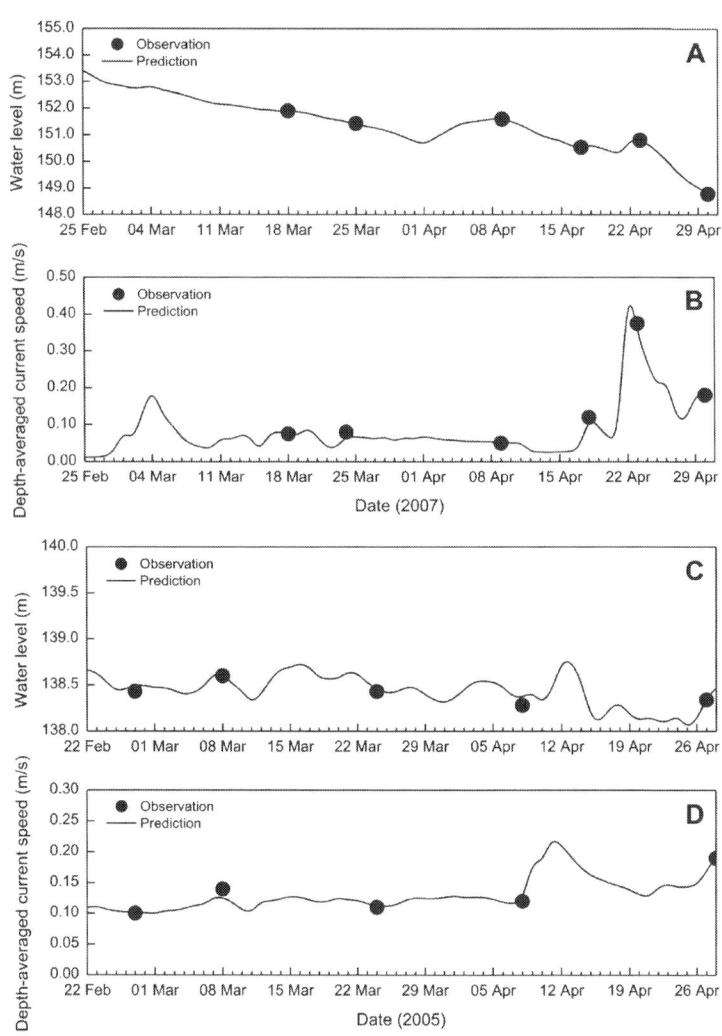

Figure 5: Comparison of observed and predicted water levels and depth-averaged current speeds: (a) and (b) at Station Zhengjiahe; (c) and (d) at Station Xiangxi mouth.

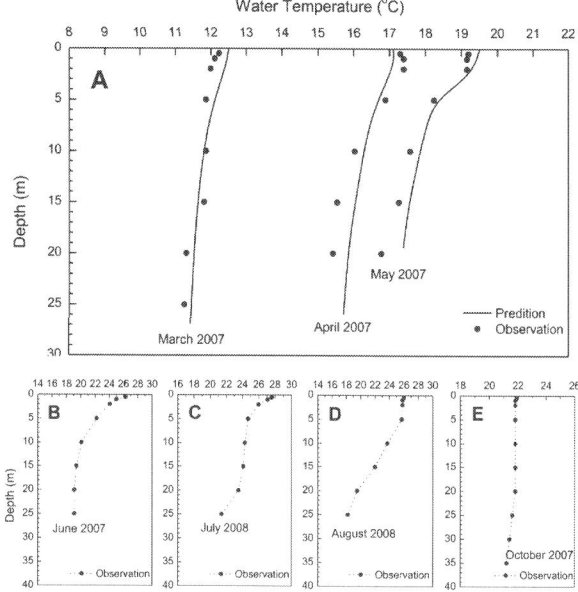

Figure 6: Temperature profiles at Station Xiakou: (a) comparisons between simulation and measurement in March, April, and May 2007; (b) observed values in June 2007; (c) and (d) observed values in July and August 2008 (after Ref [34]); (e) observed values in October 2007.

In summary, the model results could be accepted as a good approximation of actual hydrodynamic conditions. Note that there are some slight discrepancies for temperature simulation at some portions. As the generation of thermal stratification is caused by surface heat fluxes and weak vertical mixing, the discrepancies may be due to the inaccurate estimation of heat fluxes – e.g., caused by the quality of meteorological data or caused by the influence of turbidity on penetrative shortwave radiation. Also, it may be caused by the finite number of vertical layers in the model.

Longitudinal Hydrodynamic Characteristics of Xiangxi Bay

With the water level reaching over 135 m, the backwater could affect the embayment up to Xiakou Town (~25 km away from the mouth). The seasonal water temperature profiles are illustrated in Fig. 6 based on the measurements [15], [34]. It is shown that the bay remains weakly thermally stratified in the spring season ($\Delta T<3.0°C$ between surface and bottom waters), substantially stratified during the summer ($\Delta T>6.0°C$), and mixed in the autumn.

In general, the magnitude of currents is small in the bay. Fig. 7 presents the distributions of transversely-averaged flow rates in surface, middle and bottom layers for 2 representative discharges: Case I for a relatively low inflow on 11 March 2005 ($Q = 11.73$ m³/s), and Case II for a relatively high inflow on 12 April 2005 ($Q = 65.1$ m³/s). Consequently, Case II has relatively high flow rates than Case I; the maximum velocities occur at the upstream reaches and are about 0.05 m/s for Case I and 0.25 m/s for Case II, respectively.

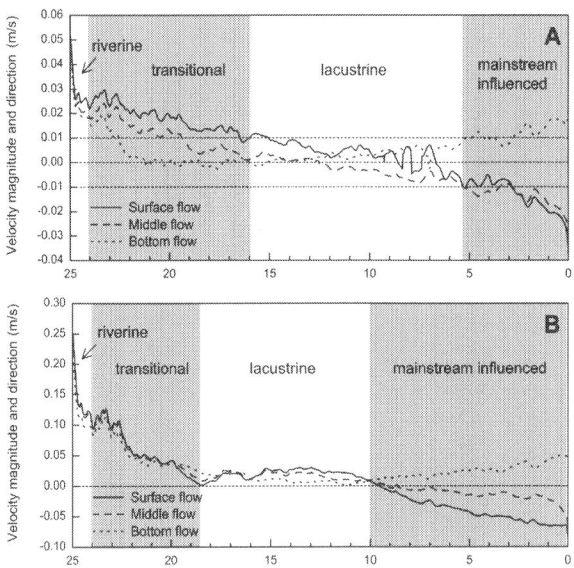

Figure 7: Computed surface, middle and bottom currents for Xiangxi Bay on (a) 11 March and (b) 12 April 2005, respectively.

Both cases exhibit a general decreasing trend of flow rates in the upstream portion, a quasi-stagnant flow in the middle portion, and a shear flow in the downstream portion. This flow regime is a response to the inflow discharges, the effects of reservoir mainstream and the vertical water temperature differences. In the upper reach, the circulation patterns are dominated by inflow-induced currents. However, the lower part is strongly affected by the reservoir mainstream. For example, in the lower part of the bay, the mainstream water enters the bay in the mid and upper layers through the mouth, while the embayment water flows out the mouth in the bottom layer (Fig. 8). The computed shear flow phenomena agree well with the measured vertical velocity patterns provided by Ref [34]. The slow flow formed in the middle reach stems from the interaction between the upstream and downstream effects.

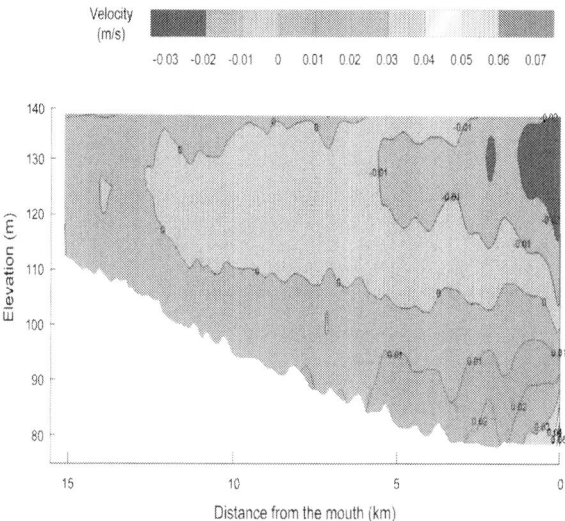

Figure 8: Transversely averaged velocities in the lower part of the bay: simulation on 11 March 2005 (top); observations on 15 March 2008 (middle) and 12 April 2008 (bottom).

The maximum influence range of the Yangtze River can extend to about 10 km upstream from the mouth (Fig. 7 and 8). This is partly because Xiangxi River has a very low flow when compared to Yangtze River; for example, during the study period from 22 February to 28 April 2005, the upstream discharges of Xiangxi River ranged from 8.17 to 99.2 m³/s, while the discharges of the reservoir mainstream were about 4800–10800 m³/s (on average 1:300). Because the backwater can affect Xiangxi River up to around 25 km from the mouth after the impoundment of TGR, the relative contribution of the upstream inflow to the overall water movement in the tributary decreases significantly with increasing wetted cross-sectional areas. Meanwhile, the weak thermal stratification in the bay can further inhibit the vertical motion, while not restricting the horizontal movement. For example, assuming that there is a uniform temperature distribution, the model results show that only the waterbodies at the mouth (~1 km) are directly controlled by the mainstream.

DISCUSSION

Longitudinal Zonation of Hydrodynamics

As stated above, a purely river-like reservoir formed in the mainstream is often divided into three longitudinal zones (Fig. 2a). Their hydrodynamic differences lead to potentially longitudinal gradients in various components (Table 1). Such a type of zonation is caused by the dam constructed at the end of the reservoir mainstream. However, unlike a purely river-like reservoir, a tributary embayment such as Xiangxi Bay is not directly obstructed by a dam, but is strongly disturbed by the outer water (reservoir mainstream). Therefore, the longitudinal zonation of a tributary embayment would differ from the typical zonation of a purely river-like reservoir. Here, we may distinguish the longitudinal zonation of Xiangxi Bay based on the hydrodynamic modeling results. As can be seen from Fig. 7, there are nonnegligible hydrodynamic gradients along the tributary embayment: (*i*) substances in the upstream part are primarily advectively controlled (steep velocity gradient); (*ii*) the importance of advection is gradually reduced in the middle-upper reach (intermediate velocity gradient), and disappears in the middle reach (negligible velocity gradient); (*iii*) the currents close to the bay mouth tend to be stratified (Fig. 7), indicating that the hydrodynamics in the downstream part is directly controlled by the reservoir mainstream. Meanwhile, for some cases of small discharges, the longitudinal gradient variations from the middle-upper reach to the middle reach are not straightforward to quantify, because the flow speeds are fairly limited (~0.01 m/s). As Ref [12] has propose a critical flow velocity (0.01 m/s) for lacustrine zone, it is possible to analyze the longitudinal zonation by combining longitudinal gradient variations (flow regime) with the critical flow velocity (additional criterion) if necessary.

Table 1: A comparison of the general properties of longitudinal zonation between normal river-like reservoirs and tributary embayments

Properties	Normal river-like reservoirs			
	Riverine	**Transitional**	**Lacustrine**	**Mainstream influenced**
Location	Upstream	Midstream	Downstream	None
Width	Narrow	Intermediate	Broad	None
Depth	Shallow	Intermediate	Deep	None
Hydrodynamics	High flow	Reduced flow	Little flow	None
Turbidity	Turbid	Less turbid	Clean	None
Trophic state	More eutrophic	Intermediate	More oligotrophic	None
Productivity	Low	Intermediate	High	None
Biodiversity	High	Intermediate	Low	None
Properties	Tributary embayments			
	Riverine	Transitional	Lacustrine	Mainstream influenced
Location	Upstream	Middle-upper	Middle-lower	Mouth
Width	Narrow	Relatively broad	Relatively broad	Relatively broad
Depth	Shallow	Relatively deep	Relatively deep	Deep
Hydrodynamics	High flow	Reduced flow	Little flow	Relatively high, stratified
Turbidity	Turbid	Less turbid	Relatively clean	Relatively clean
Trophic state	More eutrophic	Intermediate	More oligotrophic	More oligotrophic
Productivity	Low	High	High	Low
Biodiversity	High	Intermediate	Low	Low

doi:10.1371/journal.pone.0068186.t001

Fig. 9 illustrates the method how the longitudinal zones of a tributary embayment are determined, based on the characteristics of the surface, middle and bottom currents (e.g., Fig. 7). The critical velocity (e.g., absolute value of 0.01 m/s in a layer) may be used to distinguish the lacustrine zone if necessary, specifically for the cases of low inflows (e.g., Fig. 7a). Then, other zones are defined sequentially as follows: the upstream riverine zone is characterized with a relatively steep velocity gradient (~10 cm/s per km); the mainstream influenced zone showing a shear flow is close to the mouth; the velocity gradients are again employed to define the transitional zone and the lacustrine zone (for the cases of high inflows, e.g., Fig. 7b). Table 1 lists the general properties of different zones between purely river-like reservoirs and tributary embayments. Their geographical similarities and differences are also graphically shown by Fig. 2. It should be noted that, longitudinal zonation can vary greatly for different inflow discharges in a given reservoir tributary embayment (Fig. 7): for the low inflow, the lacustrine and transitional zones may occupy relatively large portions of the bay; however, due to the strong influence of the mainstream for a large discharge, the extended mainstream influenced zone can compress the space of lacustrine zone and force it to move upstream. For the purposes of regional environmental management, it is more convenient to define a quasi-deterministic zonation of hydrodynamics of Xiangxi Bay. Combining the hydrodynamic model and the longitudinal zonation method above-mentioned, Fig. 10 gives a management-oriented map of longitudinal zonation on the averaged zone boundaries over all simulations using different boundary conditions. It provides a general framework to interpret the complex spatial and temporal hydrodynamics of tributary embayments, and thus may be easily used by various stakeholders. For instance, it can be observed that under different water levels after the impoundment (>135 m), the range of mainstream influenced zone is almost limited in a 7 km reach closed to the mouth. However, the lacustrine zone continues to expand as the result of increased water levels and, at the same time, the intermediate zone gradually moves upstream.

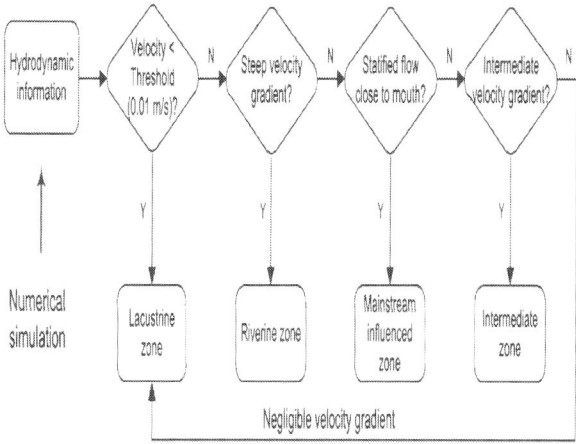

Figure 9: A schematic method of determining the longitudinal zones for a tributary embayment.

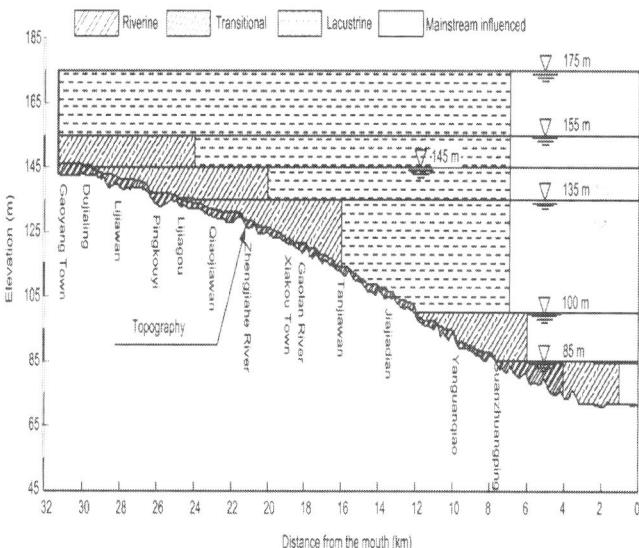

Figure 10: Map of longitudinal zonation under different water levels of Xiangxi Bay.

Effects of Hydrodynamic Characteristics on Algal Bloom Risks

Normally, an algal bloom is detected when chlorophyll-a (Chla) concentrations exceed a threshold (e.g., 10–12 µg/L), when cell counts are in the order of 10^3–10^4/mL [35]. It is interesting to estimate the algal bloom risks for different longitudinal zones, which have an early diagnostic significance in reservoir environmental management. The spatial-temporal distributions of water quality parameters that are of relevance for the algal blooms during Spring 2005 are first illustrated for several key monitoring stations (Fig. 11), of which the sites of Zhengjiahe (Z) and Xiakou (X) stand for the transitional zone, and Jiajiadian (J) and Guanzhuangping (G) represent the lacustrine zone (Fig. 1). Then, to quantify the importance of flow speeds in regulating the algal biomass, the relationship between important environmental factors and chlorophyll-a (Chla) is examined, using the data collected from Xiangxi Bay during Spring 2005. The method of regression analysis is used, and the differences between the relevant data sets are evaluated by unpaired t-tests.

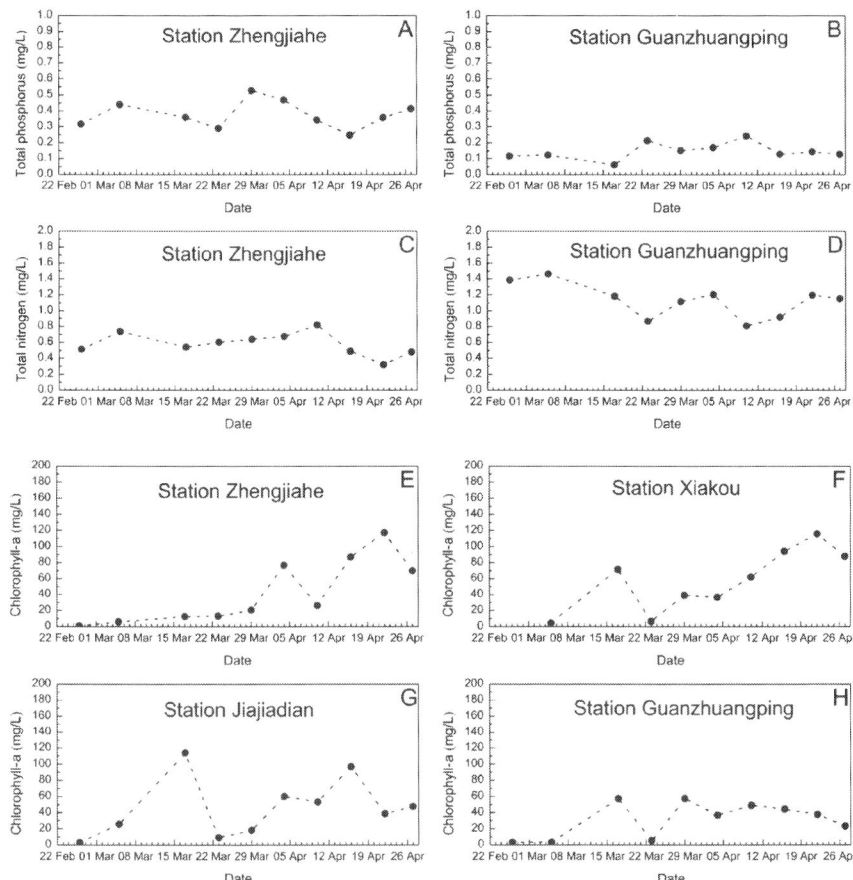

Figure 11: Observed water quality of TP, TN and Chla in Xiangxi Bay during Spring 2005.

The nutrient levels (Fig. 11a–d) indicate that the study area is in a eutrophic condition according to traditional lake trophic status indices. The observed total phosphorus (TP) concentrations are very high due to excessive phosphorus loading from the upper basin, ranging between 0.25 and 0.53 mg/L at station Z and between 0.06 and 0.24 mg/L at Station G, respectively. The mean TP value significantly exceeds the eutrophication threshold of 0.02 mg/L recommended by Ref [36]. The average total nitrogen (TN) levels are around 0.58 mg/L at station Z and around 1.13 mg/L at station G, respectively, further suggesting that the bay is suffering from excessive nutrients [37]. Therefore, algae species have a big chance to grow and bloom in the whole eutrophic area under favorable temperature, sunlight and hydrodynamic conditions.

Across the observations at Station Z and X and J (Fig. 11e–g), there is a similar temporal pattern of the Chla concentrations: (*i*) around mid-March, the primary peak originally occurred within the domain from Xiakou to Jiajiadian; the peak concentration was about 80–120 µg/L for the middle section and decreases gradually downstream; afterwards, possibly owing to the sleet weather conditions on 12 March and during 19–22 March, this bloom disappeared quickly and did not spread to the upstream section; (*ii*) the second bloom originally occurred within the middle reach around 4 April, with the maximum value of about 80 µg/L at Station Z; (*iii*) after 2 weeks, there was another extensive bloom that occurred in the same domain, with the peak value as high as 117 µg/L; the Chla level in the lower reach was dramatically decreased (Fig. 11h), which was partially due to the relatively rapid water exchange with the reservoir mainstream outside; the bloom eventually disappeared after a week of heavy rainfall. The spatial and temporal distribution of Chla concentrations shows that, in general, there is a gradually decreasing tendency for Chla along the middle and lower bay, with the higher values occurring in the places over 15 km from the mouth (Fig. 12). In particular during the algal bloom periods, the longitudinal pattern of Chla concentrations is fairly consistent with the hydrodynamic zones determined.

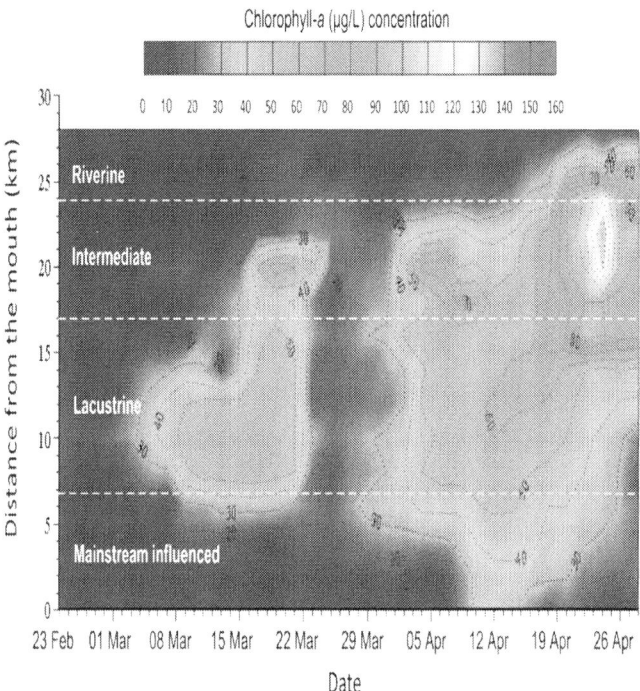

Figure 12: Observed spatial-temporal variations of chlorophyll concentrations for Xiangxi Bay during Spring 2005.

The observed total nitrogen (TN):total phosphorus (TP) ratio varies between 0.6 and 18.9 but is generally less than 15, indicating the study area is nitrogen limited – i.e., nitrogen is often in relatively short supply compared to phosphorus. However, though there is a statistically significant relationship between log(TN) and log(Chla) (p<0.0001), log(TN) can explain only 25.5% of the variance in log(Chla) (Fig. 13c). There is no significant relationship between log(TP) and log(Chla) (p = 0.087), and the former can explain only 1.5% of the variance in the latter (Fig. 13d). Moreover, a multiple regression analysis with both TN and TP can only modestly improve the correlation with TN alone [log(Chla) = 11.98–2.7 log(TN)–1.24 log(TP), with r^2 = 0.34]. Similarly, the relationship of water temperature (WT)-Chla is relatively significant (p = 0.001), while WT can only explain 14.5% of the variance in Chla (Fig. 13e). Analyzing the observed chlorophyll-*a* concentrations and computed flow velocities (U) during Spring 2005, it is found that almost all marked blooms (e.g., Chla >40 µg/L) are accompanied by very low flow rates less than 0.01 m/s (Fig. 13a, b), which also serves as a cross-check of the validity of the critical speed adopted for the longitudinal zonation method. Chlorophyll-*a* concentrations can sometimes still remain at low levels in the low flow environment, indicating that development of blooms may be suppressed by some external forces such as water temperature and light. When the flow speed is greater than the value of 0.01 m/s, there is a significant inverse relation between phytoplankton biomass and flow rates (Chla = $1.188U^{-0.938}$, with r^2 = 0.51 and P = 0.001). Therefore, under the conditions of very high nutrient concentrations, flow velocity is considered to be the primary factor regulating chlorophyll levels.

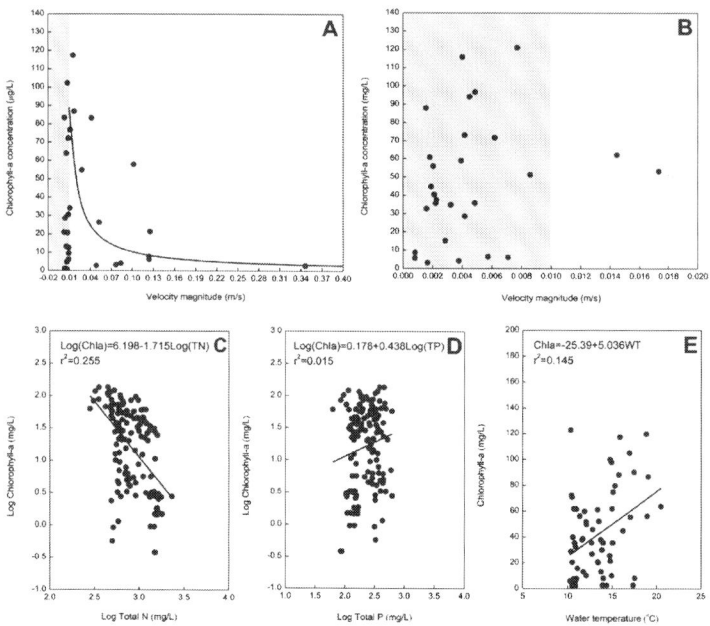

Figure 13: Scatter plot between the flow rates and chlorophyll concentrations for (a) the reach of 20–26 km and (b) the reach of 10–16 km from the mouth; (c) relationship between log total N vs. log chlorophyll; (d) log total P vs. log chlorophyll; (e) water temperature vs. chlorophyll.

Furthermore, we investigate all statistical locations where the algal blooms were initially observed during 2003–2009. Most blooms,

68.3%, first occurred in the intermediate or lacustrine zones; the lacustrine zone has the highest risk, 42.1% (Table 2). In particular, the reach of about 10 km long around Xiakou Town and Gaolan River is the most sensitive area. Only two bloom events were observed in the mainstream influenced zone during early June 2003 and late July 2004, respectively. Note that no algal blooms were observed in the riverine zone, indicating the very low risk of blooms for the upstream area. The analysis result suggests that algal blooms are prone to first occur in the very weakly flushed middle section though the whole embayment has been eutrophic. In other words, for a eutrophic bay, nutrient availability perhaps is not the exclusive causative factor for the algal blooms in the middle section, while the longitudinal hydrodynamic pattern plays a relatively major role in determining the high risk zones.

Table 2: Initial locations and biomass of algal blooms observed at Xiangxi Bay during 2003–2009

Date	Initial location	Algal concentration 110^7 cell/L)	Chla (mg/L)	Source
5-10 Jun 2003	Mainstream influenced zone	1.3		Ref [13]
Feb 2004	Whole bay	4.95		Ref [2]
Feb 2004	Transitional zone	4.7-5.2		Ref [2]
Apr 2004	Lacustrine zone	31.5		Ref [2]
Jun 2004	Transitional zone	5.8-9.2		Ref [2]
Late Jul 2004	Mainstream influenced zone	4000		Ref [2]
4-12 Mar 2005	Transitional zone		50-60	In situ observation
Apr 2005	Whole bay		60-120	In situ observation
Jun-Jul 2005	Transitional zone		60-100	Ref [13]
25 Feb 2007	Transitional zone		158	Ref [38]
25 Mar-9 Apr 2007	Whole bay		90-302	Ref [38]
6-15 May 2007	Whole bay		55	Ref [38]
Feb-Mar 2008	Lacustrine zone		25-200	CTG-TEPD
Jun-Aug 2008	Lacustrine zone	3.93-25.8		CTG-TEPD
21-28 Sep 2008	Lacustrine zone		30-60	Ref [26]
18-22 Oct 2008	Lacustrine zone		15-35	Ref [26]

10-24 Feb 2009	Lacustrine zone		1-20	*In situ* observation
13-27 Mar 2009	Lacustrine zone		20-55	*In situ* observation
11-28 Apr 2009	Lacustrine zone	-	20-35	*In situ* observation

CONCLUSIONS

This study focuses on the longitudinal hydrodynamic characteristics and their effects on algal blooms in the tributary embayment of Xiangxi Bay. The spatial and temporal distributions of hydrodynamics are successfully simulated. The results show that although the currents are generally weak, there are distinct hydrodynamic gradients within the embayment. A method of determining the bounds of longitudinal zones within the tributary embayment is then developed. A tributary embayment can be generally characterized as the riverine, intermediate, lacustrine, and mainstream influenced zones. Compared to purely river-like reservoirs, there is an additional mainstream influenced zone near the mouth. The interaction between inflow discharges and the reservoir mainstream causes the existence and change of longitudinal characteristics.

The physical differences of the longitudinal zones can provide insights into algal bloom initiation in tributary embayments. The high risk zones of algal blooms are characterized by slow flow and weak gradients. The lacustrine zone represents about 42.1% and the transition zone represents about 26.3% of the observed blooms, respectively. Nevertheless, note that while there is correlation between water velocity and algal growth, there is still a multitude of factors that are believed to strongly influence algal growth, including nutrient loading, solar radiation, water temperature, water level, etc.

ACKNOWLEDGMENTS

The authors would like to thank Dr. Zhenzhen Yu and Dr. Tiegang Zheng for valuable discussions and advice with the simulation experiments. We thank Dr. Yahong Dong for reviews and comments of an early copy

Longitudinal Hydrodynamic Characteristics in Reservoir Tributary... 151

of this manuscript. This manuscript also benefited from the fruitful comments from the reviewers, namely, Dr. Hong-Yi Li, Dr. Huan Wu, and an anonymous reviewer.

REFERENCES

1. Dai HC, Cao GJ, Su HZ (2006) Management and construction of the Three Gorges Project. Journal of Construction Engineering and Management-ASCE 132: 615–619. doi: 10.1061/(asce)0733-9364(2006)132:6(615)

2. Cai QH, Hu ZY (2006) Studies on eutrophication problem and control strategy in the Three Gorges Reservoir. Acta Hydrobiologica Sinica 31: 7–11.

3. Yang GS, Weng L, Li L (2007) Yangtze conservation and development report 2007. Science Press, Beijing, China.

4. Anderson DM, Glibert PM, Burkholder JM (2002) Harmful algal blooms and eutrophication: nutrient sources, composition, and consequences. Estuaries 25: 704–726. doi: 10.1007/bf02804901

5. Schindler DW, Fee EJ (1974) Experimental lakes area; whole lakes experiment in eutrophication. Journal of the Fisheries Research Board of Canada 31: 937–953. doi: 10.1139/f74-110

6. Wong KTM, Lee JHW, Hodgkiss IJ (2007) A simple model for forecast of coastal algal blooms. Estuarine, Coastal and Shelf Science 74: 175–196. doi: 10.1016/j.ecss.2007.04.012

7. Vanotte RL, Minshall GW, Cummins KW, Sedell JR, Cushing CE (1980) The river continuum concept. Canadian Journal of Fisheries and Aquatic Sciences 37: 130–137. doi: 10.1139/f80-017

8. Wang SH, Dzialowski AR, Meyer JO, Jr FD, Lim NC, et al. (2005) Relationships between cyanobacterial production and the physical and chemical properties of a Midwestern Reservoir, USA. Hydrobiologia 541: 29–43. doi: 10.1007/s10750-004-4665-x

9. Lindim C, Pinho JL, Vieira JMP (2011) Analysis of spatial and temporal patterns in a large reservoir using water quality and hydrodynamic modeling. Ecological Modelling 222: 2485–2494. doi: 10.1016/j.ecolmodel.2010.07.019

10. Kimmel BL, Groeger AW (1984) Factors controlling primary production in lakes and reservoirs: a perspective. Lake and Reservoir Management 1: 277–281. doi: 10.1080/07438148409354524

11. Thornton KW, Kimmel BL, Payne FE (1990) Reservoir limnology: ecological perspectives. New Jersey: John Wiley & Sons, Inc.

12. Ministry of Environmental Protection of China (MEPC) (2010) Technical guideline for water environmental quality assessment of Three Gorges Reservoir. Beijing, China.

13. Ye L (2006) The Occurrence rule and countermeasure study on algae blooms of Xiangxi River in the Three Georges Reservoir Area. Master's thesis. Hohai University, China.

14. Ye L, Han XQ, Xu YY, Cai QH (2007) Spatial analysis for spring bloom and nutrient limitation in Xiangxi Bay of Three Gorges Reservoir. Environmental Monitoring and Assessment 127: 135–145. doi: 10.1007/s10661-006-9267-9

15. Zheng TG, Mao JQ, Dai HC, Liu DF (2011) Impacts of water release operations on algal blooms in a tributary bay of Three Gorges Reservoir. Science China Technological Sciences 54: 1588–1598. doi: 10.1007/s11431-011-4371-7

16. Wang L, Cai QH, Zhang ML, Tan L, Kong LH (2010) Longitudinal patterns of phytoplankton distribution in a tributary bay under reservoir operation. Quaternary International 244: 280–288. doi: 10.1016/j.quaint.2010.09.012

17. Shao ML, Xu YY, Cai QH (2010) Effects of reservoir mainstream on longitudinal zonation in reservoir bays. Journal of Freshwater Ecology 25: 107–117. doi: 10.1080/02705060.2010.9664363

18. Wang LL, Yu ZZ, Dai HC, Cai QH (2009) Eutrophication model for river-type reservoir tributaries and its applications. Water Science and Engineering 2: 16–24.

19. Sheng YP (1987) On modeling three-dimensional estuarine and marine hydrodynamics. In: Nihoul, Jamart (Eds.), Three-dimensional Marine and Estuarine Hydrodynamics. Elsevier, Amsterdam: 35–54.

20. Delft Hydraulics (2006) Delft3D-FLOW user manual, Delft, the Netherlands.

21. Hamrick JM (1996) A user's manual for the environmental fluid dynamics computer code (EFDC). Special Report 331. The College of William and Mary, Virginia Institute of Marine Science.

22. Chen C, Liu H, Beardsley RC (2003) An unstructured, finite-volume, three-dimensional, primitive equation ocean model: application to coastal ocean and estuaries. Journal of Atmospheric and Oceanic Technology 20: 159–186. doi: 10.1175/1520-0426(2003)020<0159:augfvt>2.0.co;2

23. Danish Hydraulic Institute (DHI) (2007) Mike Zero: The common DHI user interface for project oriented water modeling. DHI Water & Environment: Denmark.

24. Shchepetkin AF, McWilliams JC (2005) The Regional Ocean Modeling System (ROMS): A split-explicit, free-surface, topography following coordinates ocean model, Ocean Modelling. 9: 347–404. doi: 10.1016/j.ocemod.2004.08.002

25. Cole RW, Buchak EM (1995) CE-QUAL-W2: A two dimensional, laterally averaged, hydrodynamic and water quality model. Version 2.0. Instruction Rep. EL-95-1, U.S Army Engineer Waterways Experiment Station, Vicksburg, Miss.

26. Yang ZJ, Liu DF, Ji DB, Xiao SB (2010) Influence of the impounding process of the Three Gorges Reservoir up to water level 172.5 m on water eutrophication in the Xiangxi Bay. Science China Technological Sciences 53: 1114–1125. doi: 10.1007/s11431-009-0387-7

27. van Maren DS, Winterwerp JC, Wu BS, Zhou JJ (2009) Modelling hyperconcentrated flow in the Yellow River. Earth Surf Process Landforms 34: 596–612. doi: 10.1002/esp.1760

28. Chanudet V, Fabre V, van der Kaaij T (2012) Application of a three-dimensional hydrodynamic model to the Nam Theun 2 Reservoir (Lao PDR). Journal of Great Lakes Research 38: 260–269. doi: 10.1016/j.jglr.2012.01.008

29. Lee JHW, Qu B (2004) Hydrodynamic tracking of the massive spring 1998 red tide in Hong Kong. Journal of Environmental Engineering-ASCE 130: 535–550. doi: 10.1061/(asce)0733-9372(2004)130:5(535)

30. Choi KW, Mao JQ, Lee JHW (2007) Technical Note on Validation of the 3D flow and mass transport model (EFDC/DESA). Aoe-water website. Available: http://www.aoe-water.hku.hk/visjet/itsp2006.htm. Accessed 2013 Jun 17.

31. .Dong YH (2011) Analysis of stratification and algal bloom risk in Mirs Bay. Master's thesis. The University of Hong Kong, Hong Kong.

32. 32.Mao JQ, Wong KTM, Lee JHW, Choi KW (2011) Flushing time of marine fish culture zones in Hong Kong. China Ocean Engineering 25: 625–643. doi: 10.1007/s13344-011-0050-5

33. Rodi W (1980) Turbulence models and their application in hydraulics - a state of the art review. IAHR monograph, Delft, Netherlands.

34. Yi ZQ, Liu DF, Yang ZJ, Ma J, Ji DB (2009) Water temperature structure and impact of which on the bloom in spring in Xiangxi Bay at Three Gorges Reservoir. Journal of Hydroecology 2: 6–11.

35. .Mao JQ, Lee JHW, Choi KW (2009) The extended Kalman filter for forecast of algal bloom dynamics. Water Research 43: 4214–4224. doi: 10.1016/j.watres.2009.06.012

36. .United States Environmental Protection Agency (USEPA) (1974) An approach to a relative trophic index system for classifying lakes and reservoirs. National Eutrophication Survey Working Paper No.24.

37. .Vollenweider RA (1968) Scientific fundamentals of the eutrophication of lakes and flowing waters with special reference to nitrogen and phosphorus as factors in eutrophication, Technical Report. Organisation for Economic Co-operation and Development, Paris.

38. Huang YL (2007) Study on the formation and disappearance mechanism of algal bloom in the Xiangxi River at Three Gorges Reservoir. Master's thesis. Northwest A&F University, China.

Chapter 7

Vertical Variation of Nonpoint Source Pollutants in the Three Gorges Reservoir Region

Zhenyao Shen, Lei Chen, Qian Hong Hui Xie, Jiali Qiu, Ruimin Liu

State Key Laboratory of Water Environment Simulation, School of Environment, Beijing Normal University, Beijing, P.R. China

ABSTRACT

Nonpoint source (NPS) pollution is considered the main reason for water deterioration, but there has been no attempt to incorporate vertical variations of NPS pollution into watershed management, especially in mountainous areas. In this study, the vertical variations of pollutant yields were explored in the Three Gorges Reservoir Region (TGRR) and the relationships between topographic attributes and pollutant yields were established. Based on our results, the pollutant yields decreased significantly from low altitude to median altitude and leveled off rapidly from median altitude to high altitude, indicating logarithmic relationships between pollutant yields and altitudes. The pollutant yields peaked at an altitude of 200–500 m, where agricultural land and gentle slopes (0–8°) are concentrated. Unlike the horizontal distributions, these vertical variations were not always related to precipitation patterns but did vary obviously with land uses and slopes. This paper also indicates that altitude data and proportions of land use could be a reliable estimate of NPS yields at different altitudes, with significant implications for land use planning and watershed management.

INTRODUCTION

After decades of working to reduce emissions from point sources, problems regarding nonpoint source (NPS) pollution have been highlighted, with agriculture being the largest contributor [1]. The three main forms of NPS pollutants are sediments, nutrients and pesticides [2], the effects of which are well documented [3]–[5]. Researchers have revealed that NPS pollution may come from a wide range of dispersed sources though a complex combination of physical, chemical and biological processes [6]. From an environmental point of view, there is a dire need to gain insights into the spatial variations of NPS pollution, which are essential for analyzing these complex problems in drainage basins.

In general, the spatial distributions of NPS pollution can be quantified by monitoring or modeling methods. In respect to monitoring strategy, detailed measured data are collected and the spatial variations of water quality can be analyzed by comparing those measured data. However, water quality degradation often results from multiple sources and separating the impacts by monitoring methods is very difficult and costly, especially for a large basin [7]. Watershed models can facilitate in identifying individual sources of NPS pollution and evaluating the decision schemes for watershed management. Up to now, many models have been developed for identifying the spatial distributions of NPS pollution [8]. Some of these models, such as Export Coefficient Model [9], are based on empirical equations and cannot always provide sufficient explanations for those complex watershed processes [10]. By contrast, those physically-based models can simulate the hydrologic and water quality responses at varying scopes and locations [11]. In addition, those physically-based models are usually coupled with the geographic information system (GIS) which can compile extensive input database and visualize the model results. Information of watershed characteristics can be extracted and analyzed with convenience of GIS techniques which are usually integrated into these watershed models. The most commonly-used watershed models are the Soil and Water Assessment Tool (SWAT) model [12], Agricultural Nonpoint Source pollution (AGNPS) model[13], Annualized Agricultural Nonpoint Source pollution (AnnAGNPS) model [14], and Hydrological Simulation Program - Fortran (HSPF) [15].

Currently, the GIS techniques have provided a reliable platform for integrating vertical and horizontal information within a basin. Within such a framework, altitude data are generally extracted from a Digital Elevation Model (DEM) [16], and this 3-dimensional information has been widely applied in studies on atmospheric pollution [17], [18]. However, there is current interest in integrating the GIS platform to project large volumes of meteorological and geophysical data into horizontal information to study NPS pollution [19], regardless of whether vertical variations occur. Indeed, altitude is the key attribute of topography and has a direct impact on physical parameters such as precipitation, solar radiation, temperature and soil chemistry [20], [21]. Altitude is also essential in other environmental factors, including slope length, slope degree and other properties [22], [23]. Researchers have reported that altitude has an impact on geomorphologic processes such as surface runoff, soil erosion and landslides in hilly regions [24], [25]. Land use changes, landscape dynamics and other human activities are therefore related to the terrain and altitude [26], especially in the mountainous areas. Therefore, those projected horizontal distributions of NPS pollution is a consequence but may not be the cause of NPS pollution because this information should make reference to specific vertical processes. As far as we know, there has been no attempt to incorporate vertical variations for the analysis of NPS pollution, which should draw increasing attention due to continued hilly urbanization, increased deforestation, and changed precipitation with global warming [27].

The objective of this paper is to contribute new insights into vertical variations to capture the complex features of NPS pollution. The study was performed in the Three Gorges Reservoir Region (TGRR) by: 1) exploring the spatial distributions of sediment, nitrogen (N), and phosphorus (P) using the Soil and Water Assessment Tool (SWAT); 2) establishing the relationships between land use, slope and altitude; and 3) characterizing the vertical variations of sediment, N and P yields in the TGRR.

MATERIALS AND METHODS

Watershed Description

The Three Gorges Reservoir, which is by far the world's largest hydropower project, completed its first filling stage in 2003 and reached its maximum designed water level in 2008. Geographically, the TGRR, with a total area of approximately 58,000 km^2, is located in the transitional zone from the Tibetan Plateau in the west to the east rolling hills and plains of China between latitudes 28°10′ and 32°13′N and longitudes 105°17′ and 110°11′E (Fig. 1). The topography is complex, with over 74% of the landscape being mountainous and 21.7% being low hills. The land uses include cropland (39%), grassland (13%) and forest (46%), while the main soils are purplish soils (48%), limestone earths (34%) and yellow (16%) earths. The average precipitation is approximately 1400 mm, 80% of which occurs from April to October. The highest and lowest annual temperature ranges from approximately 27°C to 29°C and 6°C to 8°C, respectively.

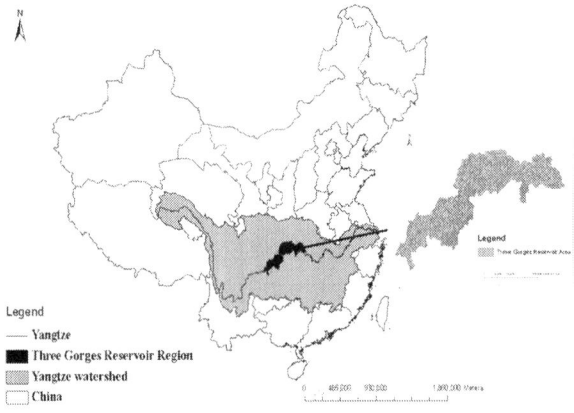

Figure 1: The location of the Three Gorges Reservoir Region.

When water levels were driven up by the Three Gorges Reservoir, hilly reclamation and deforestation continued to increase above the 175-m inundation line [28], [29]. Due to the special geography and structure of the agriculture in the TGRR, the soil loss is serious and the eco-environment is vulnerable. Additionally, after the water was cut off, the water velocity was reduced and the retention time of pollutants prolonged. The water quality challenge has never been greater than now, as indicated by the soil erosion in the uplands and algal blooms in the aquatic environment [29].

Model Description and Preparation

Model Description

The ArcSWAT model, developed by Arnold et al. [12], was used to develop the necessary input files. The SWAT components include weather generation, hydrology, soil erosion, crop growth, nutrient leaching and agricultural management [30]. The hydrology calculation was based on the curve number method and the Green-Ampt infiltration method [31]. The sediment yield was estimated by the modified soil loss equation [32]. Runoff, sediments and nutrients were calculated for each Hydrologic Response Unit (HRU) and then routed in stream using the QUAL2E model [33]. More information about the SWAT model can be found in Douglas et al.[30] (Methods S1).

Data Description

Taking into account the study needs and data availability, the digital layers for altitude, land use and soil were constructed. DEM data at a scale of 1:25,000 published by the Institute of Geographical and Natural Resources Research, China, were used. Land-use data were interpreted from a 1:100,000 Thematic Mapper image and the proportions of land uses were treated as constants during the simulation period. A soil map at a scale of 1:1,000,000 and the related physical data were obtained from the Institute of Soil Science, Chinese Academy of Sciences. The daily precipitation, relative humidity, solar radiation, wind speed and air temperature data, measured by 49 weather stations from 1980 to 2010, were obtained from the State Meteorological Data Sharing

Service System (http://cdc.cma.gov.cn). Crop information, including tillage, irrigation and the amount of fertilizer used, was based on statistical data from local bureaus as well as field investigations in several local watersheds. Rice, potato, sweet potato and corn were selected as the main crops due to the cultivated areas and the amount of fertilizer. However, no clear records of management practices were available. To compensate for this lack, the average fertilizer rates in each village were calculated for cultivated crops and all agricultural areas were assumed to be tile drained.

Model Preparation

In this study, the TGRR was delineated into 613 sub-watersheds interconnected by a stream network, and each sub-watershed was divided further into HRUs by setting 0% thresholds of land use, soil type and slope to accurately capture even small areas. In our previous study [34], we introduced a small-scale watershed extended method (SWEM) for parameter calibration in the TGRR (Methods S2). The detailed processes involve: 1) model calibration- a process of generating model parameter groups for representing different parts of the TGRR (in terms of the watersheds of the Yulin, Xiaojiang, Daning and Xiangxi); 2) extended modeling- running the well-calibrated models in the corresponding parts of the TGRR The measured flow and water quality data were obtained from the Changjiang Water Resources Commission and the parameter groups were generated by the SWAT-CUP [35]. The Nash-Sutcliffe efficiency coefficient (E_{NS}) [36] was used to quantify the degree of fit between the simulated data and the measured data.

$$E_{NS} = 1 - \frac{\sum_{i=1}^{n} \left(Q_{sim,i} - Q_{mea,i} \right)^2}{\sum_{i=1}^{n} \left(Q_{mea,i} - \overline{Q}_{mea} \right)^2}$$

(1)

Where, $Q_{mea,i}$ is the ith observation for the constituent being evaluated, $Q_{sim,i}$ is the predicted value for the constituent being evaluated, $\overline{Q}_{mea}$ is the mean value of observed data for the constituent being evaluated, and n is the total number of observations.

The values of E_{ns} in the respective sub-watersheds ranged from 0.53 to 0.94 for the stream flow, 0.53–0.94 for sediment, 0.60–0.84 for total P (TP), 0.47–0.80 for nitrate-N and 0.41–0.81 for NH_4-N. The detailed processes of the model calibration and validation can be found in our previous study [34], [37]. With the groups of calibrated parameters, the extended simulation was conducted by running the well-calibrated SWAT models in the entire TGRR.

Data Analysis

Following calibration, a 10-year (2000–2009) simulation was performed to isolate variability of climate, land use, crop rotations and runoff regime which may mask the effect of vertical variation [38]. The DEM and land use maps were divided into two matrixes of 4500 rows by 5500 columns, with a cell size of 100*100 m. A mask layer was applied to all cells to avoid noise in the statistical processing of the data, and 6,200,000 cells remained after this selection. Topographic data such as altitude, slope and land use were generated for each cell, and the flow, sediment and nutrient yield were calculated from SWAT outputs. The total yields were defined as the summary of corresponding cells.

RESULTS AND DISCUSSION

The Vertical Variations of Land Use and Slope

Fig. 2 illustrates the vertical variations of the land uses at different altitudes. According to Fig. 2, the landscape area increases significantly from 0 m to 400 m and levels off rapidly from 400 m to 1600 m, while only slight declines can be observed when the altitude varies from 1600 m to 2100 m. Specifically, the land from 200 m to 1000 m was dominant in the TGRR, covering more than 71% of the entire area. The landscape area accounted for only 2% from 0 m to 200 m, 13% from 1000 m to 1500 m and 2% from 1500 m to 2100 m of the total area. Among the different land uses, agriculture (paddy field and dry land), forest and grassland were dominant in all altitudes but the vertical variations of these land uses were different. As illustrated in Fig. 2, the proportions of agriculture show obvious declines as the altitude increases, while those of forest and grassland show increasing trends. In particular, agricultural areas were concentrated among the altitudes between 200 m and 800 m. This vertical variation could be explained by most low-altitude areas below the 175-m inundation line having been submerged when the water levels were driven up by the Three Gorges Dam [29], [38]. These vertical variations of land uses were also indicated by other studies showing that farmers resettled in the low hilly areas and 80% of the arable farmland is distributed in the low hilly areas or valley terraces [39],[40]. In this study, the slope degrees were categorized into 0–8°, 8–15°, 15–25°, 25–35° and 35~90°. The vertical variations of these slope degrees are analyzed in Fig. 3. In the TGRR, gentle slope (0–8°) made up the largest proportion (31%) of the entire area, while the land on median slope (8–25°) and steep slope (25–90°) accounted for 50% and 19%, respectively.

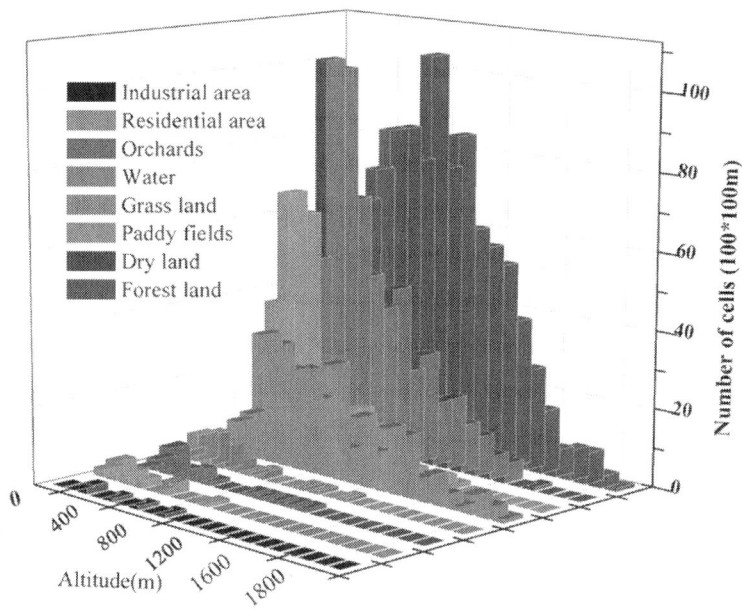

Figure 2: The vertical variations of land use.

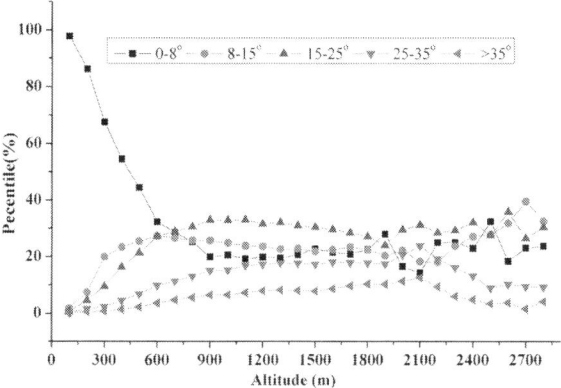

Figure 3: The vertical variations of slope.

Fig. 3 also illustrates that the vertical variations of slope degree show obvious trends. Below the altitude of 800 m, the proportions of gentle slope gradually decreased as altitude increased, while the proportions of median and steep slopes increased slightly. Between the altitudes of 800 m and 2000 m, the proportions of gentle slope and median slope remained stable, while the proportions of steep slope continued to increase. Above the altitude of 2000 m, the landscape area was again dominated by gentle and median slopes. This result can be explained by the widely held view that the rock strength at high altitudes is normally high and that such altitudes usually have weathered rocks or rocks whose shear strength is much higher[25], indicating the high altitude areas in the TGRR are generally flat and have a convergent terrain.

The Vertical Variations of NPS Pollution

Fig. 4 further characterizes the vertical variation of precipitation, sediment, N and P yields. Fig. 4 illustrates that precipitation did not vary significantly with altitude. This result is inconsistent with previous studies [16], [41] that demonstrated that precipitation increased with altitude due to the orographic effect, which lifted the air vertically and the condensation occurred due to adiabatic cooling. There are two probable reasons for this inconsistent variation. First, the climate in the TGRR is subtropical, with the annual mean temperature being 17°C, so adequate illumination may compensate for the orographic effect in the mountainous terrain. Second, a 10-year period was considered in the current study to represent the climatic variations. As rainfall is irregular in occurrence, duration and magnitude, this long period is equally true for a flattening effect of precipitation [41]. This paper indicates that NPS pollution did vary with altitude, even in the absence of different precipitation patterns related to altitude.

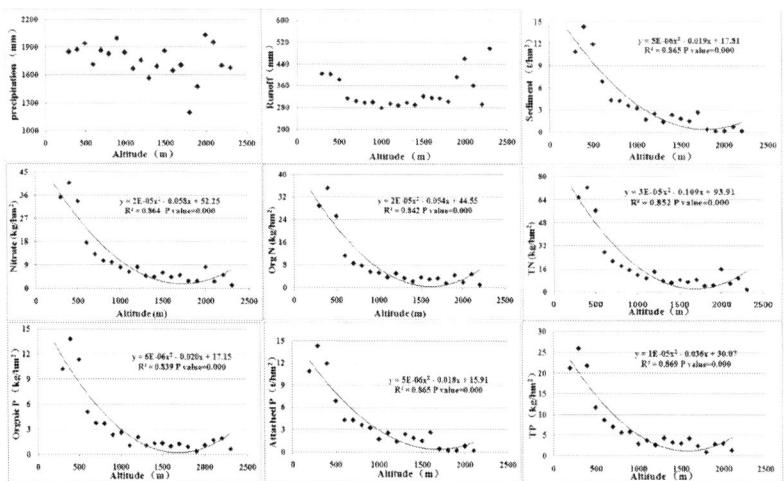

Figure 4: The vertical variations of precipitation and pollution yield.

As illustrated in Fig. 4, the load intensities of all pollutants showed obvious declines from low altitude to high altitude. All variables peaked at the low altitude (200–500 m), where frequency of human actives is the highest. Specifically, soil erosion (above 500 t/(km² a)) occurred in over 90% of the altitudes of the TGRR, while 33% of the areas were heavily eroded below the altitude of 500 m, with an erosion coefficient greater than 4,000 t/(km² a). This result can be explained by the rock strength and vegetation cohesion making the high altitudes pollutant sinks, while the low altitude areas were normally prone to environmental vulnerability due to human disturbances [24], [25]. The logarithmic lines were generated to demonstrate the correlation between the pollutant yields and altitude (Fig. 4). As shown in Fig. 4, the regression results are significant, with the regression correlations being larger than 0.74.

The relationships among land use, slope and NPS yields were also explored. As illustrated inFig. 5, the proportion of agricultural area was positively correlated with pollution yields, while that of forest was negatively correlated. For every 1% reduction in forest area, the load intensity increased by 0.01~11.34 t/km^2 for sediment, 0.15~2.83 kg/km^2 for TP and 0.40~14.00 kg/km^2 for total nitrogen (TN). The main reason for this result is that forest plants generally have a higher capability to hold and fix soil, while agricultural soil is either regularly over-fertilized or highly vulnerable to erosion [42]. In the TGRR, the agricultural area shrank at a high rate due to the Three Georges Reservoir, and there was no alternative but to rely on greater applications of fertilizer to ensure high productivity for the huge and growing population [29], [38], [43]. Specifically, the sediment yield increased slightly when the proportions of agriculture changed from 0% to 40%, and it showed a jump when the agriculture varied from 40% to 60%. This phenomenon could also be observed in P and N yields, from which the jumping points were obtained at proportions of 10% and 40%, respectively. For forest, the load intensity of sediment, TN and TP remained stable outside a relevant domain of 40%, 10% and 40%, respectively, and any change inside this proportion domain would have a greater impact on NPS yields. This phenomenon may be explained by the spatial distributions of the converted landscape pattern, which may mitigate certain discharges and may not always intensify the NPS pollution [44].

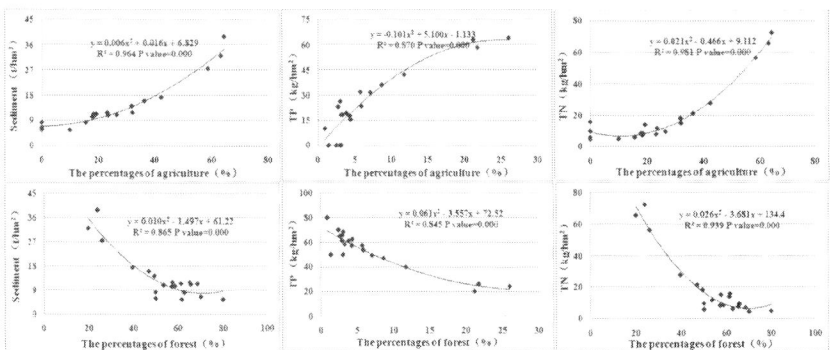

Figure 5: The relationship between pollutant yield and land uses in different altitudes.

As shown in Table 1, the load intensity of sediment, organic N and attached P progressively rises with relief up to the median slope. The pollutant yield then declines toward steep slope. The most severe pollution risks typically occurred on the slopes between 8° and 15°, which is inconsistent with the widely held view that the downhill force is highest on steeper slopes [25]. However, the TGRR is major region for national environmental protection, and many projects, such as 'Grain for Green Project', have been implemented in steep slope areas to strengthen soil and reduce NPS yields. Additionally, human practices, including rotation, irrigation and tillage, have been conducted at the gentle slope and low altitude areas; therefore, gentler slopes, particularly in deeply human-impacted slopes, may increase hydrological connectivity and nutrient leaking, resulting in greater efficiency of delivery of P and N to surface waters [42].

Table 1. The load intensities of pollutants at different altitudes

Slope	Sediment	Nitrate	Orgnic N	Attached P	Solute P	TN	TP
	t/hm²	kg/hm²	kg/hm²	kg/hm²	kg/hm²	kg/hm²	kg/hm²
0–8°	11.9	28.7	21.0	1.5	1.1	49.8	4.1
8-15°	23.0	17.0	23.4	2.0	0.6	40.5	4.1
15-25°	18.3	10.7	12.4	1.0	0.4	23.1	2.1
25-35°	15.8	5.7	6.1	0.7	0.2	11.8	1.3
>35°	16.2	3.8	3.7	0.5	0.2	7.5	0.9

CONCLUSIONS

In this paper, the vertical variations of land use, slope and NPS yields were estimated and used for studying the behavior of pollutants in the TGRR. Based on our results, the NPS pollution showed an obvious decline from low to high altitude, with all variables peaking at the low altitude (200–500 m), where the frequency of human actives was the highest. The watershed manager can gain insight into vertical dynamics to develop site-specific policies using this spatial information. This paper indicates that the vertical variations of NPS pollution were not related to precipitation patterns but did vary with vertical variations of land uses and slopes. Therefore, altitude data and proportions of land uses can be regarded as a reliable estimate of NPS load intensity, especially in the mountainous areas. However, uncertainty of modeling outcomes must be estimated to establish the reliability of the simulated outputs. In the future, more detailed data should be used and more pollutants, such as pesticide, heavy metal and Polychlorinated Biphenyl, should be incorporated into the list of analysts.

ACKNOWLEDGMENTS

The authors wish to express their gratitude to Plos One, as well as to the anonymous reviewers who helped to improve this paper though their thorough review.

AUTHOR CONTRIBUTIONS

Conceived and designed the experiments: ZYS. Performed the experiments: LC QH HX JLQ RML. Analyzed the data: ZYS LC QH HX JLQ. Contributed reagents/materials/analysis tools: ZYS LC QH HX. Wrote the paper: LC ZYS QH.

REFERENCES

1. Xu Z, Wan S, Ren H, Han X, Li M-H, et al. (2012) Effects of water andnitrogen addition on species turnover in temperate grasslands in northern china. Plos One 7: e39762.

2. Dowd BM, Press D, Los Huertos M (2008) Agricultural nonpoint source waterpollution policy: The case of California's Central Coast. Agr Ecosyst Environ 128: 151–161.

3. Nelson JL, Zavaleta ES (2012) Salt marsh as a coastal filter for the oceans:changes in function with experimental increases in nitrogen loading and sea-level rise. Plos One 7: e38558.

4. Keatley BE, Bennett EM, MacDonald GK, Taranu ZE, Gregory-Eaves I (2011) Land-Use Legacies Are Important Determinants of Lake Eutrophication in the Anthropocene. Plos One 6: e15913.

5. Somura H, Takeda I, Arnold JG, Mori Y, Jeong J, et al. (2012) Impact of suspended sediment and nutrient loading from land

uses against water quality in nthe Hii River basin, Japan. J Hydrol 450: 25–35.

6. Short JS, Ribaudo M, Horan RD, Blandford D (2012) Reforming Agricultural Nonpoint Pollution Policy in an Increasingly Budget-Constrained Environment. Environ Sci Tech 46: 1316–1325.

7. Shen ZY, Chen L, Chen T (2012) The influence of parameter distribution uncertainty on hydrological and sediment modeling: a case study of SWAT model applied to the Daning watershed of the Three Gorges Reservoir Region, China. Stochc Env Res Risk A. 27: 235–251.

8. Shen ZY, Liao Q, Hong Q, Gong YW (2011) An overview of research on agricultural non-point sources pollution modelling in China. Sep Purif Technol. 9: 595–604.

9. Ding X, Shen ZY, Hong Q, Yang ZF, Wu X, et al. (2010) Development and test of the Export Coefficient Model in the Upper Reach of the Yangtze River. J Hydrol. 383: 233–244.

10. Shen ZY, Chen L, Hong Q, Ding XW, Liu RM, et al. (2013) Long-term variation (1960–2003) and causal factors of non-point source nitrogen and phosphorus loads in the Upper Reach of Yangtze River. J Hazard Mater. 252– 253: 45–56.

11. Woznicki S, Nejadhashemi A, Smith C (2011) Assessing best management practice implementation strategies under climate change scenarios. T ASABE. 54: 171–190.

12. Arnold JG, Srinivasan R, Muttiah RS, Williams JR (1998) Large area hydrologic modeling and assessment - Part 1: Model development. J Am Water Resour As 34: 73–89.

13. Young RA, Onstad C, Bosch D, Anderson W (1989) AGNPS: A nonpointsource pollution model for evaluating agricultural watersheds. J Soil Water Conserv. 44: 168–173.

14. Bingner R, Theurer F, Yuan Y (2001) AnnAGNPS Technical Processes: Documentation Version 2. Unpublished Report, USDA-ARS National Sedimentation Laboratory, Oxford, Miss. NPS Vertical Variation in the TGRR PLOS ONE | www.plosone.org 6 August 2013 | Volume 8 | Issue 8 | e71194

15. Bicknell B, Imhoff J, Kittle Jr J, Donigian Jr A, Johanson R (1993) Hydrologic Simulation Program-FORTRAN (HSPF): User's Manual for Release 10. Rep. No. EPA/600/R-93/174. US EPA Environmental Research Lab, Athens, Ga.

16. Lin K, Zhang Q, Chen X (2010) An evaluation of impacts of DEM resolution and parameter correlation on TOPMODEL modeling uncertainty. J Hydrol 394: 370–383.

17. Moreau-Guigon E, Motelay-Massei A, Harner T, Pozo K, Diamond M, et al. (2007) Vertical and temporal distribution of persistent organic pollutants in Toronto. 1. Organochlorine pesticides. Environ Sci Tech 41: 2172–2177.

18. Hauck M, Zimmermann J, Jacob M, Dulamsuren C, Bade C, et al. (2012) Rapid recovery of stem increment in Norway spruce at reduced SO2 levels in the Harz Mountains, Germany. Environ Pollut 164: 132–141.

19. Wong MS, Nichol JE, Lee KH (2009) Modeling of Aerosol Vertical Profiles Using GIS and Remote Sensing. Sensors 9: 4380–4389.

20. Bryan BA (2003) Physical environmental modeling, visualization and query for supporting landscape planning decisions. Landscape and Urban Plan. 65: 237– 259.

21. Wu S, Li J, Huang GH (2007) Characterization and Evaluation of Elevation Data Uncertainty in Water Resources Modeling with GIS. Water Resour Manage 22: 959–972.

22. Livne E, Svoray T (2011) Components of uncertainty in primary production model: the study of DEM, classification and location error. Int J Geogr Inf Sci25: 473–488. 23. Lin K, Zhang Q, Chen X (2010) An evaluation of impacts of DEM resolution and parameter correlation on TOPMODEL modeling uncertainty. J Hydrol394: 370–383.

23. Pourghasemi HR, Mohammady M, Pradhan B (2012) Landslide susceptibility mapping using index of entropy and conditional probability models in GIS: Safarood Basin, Iran. Catena 97: 71–84.

24. Ghimire M (2011) Landslide occurrence and its relation with terrain factors in the Siwalik Hills, Nepal: case study of susceptibility assessment in three basins. Nat Hazards56: 299–320.

25. Orgiazzi A, Lumini E, Nilsson RH, Girlanda M, Vizzini A, et al. (2012) Unravelling Soil Fungal Communities from Different Mediterranean Land-Use Backgrounds. Plos One 7: e34847.

26. Guo SL, Wang JX, Xiong LH, Ying AW, Li DF (2002) A macro-scale and semidistributed monthly water balance model to predict climate change impacts in China. J Hydrol 268: 1–15.

27. Shi ZH, Ai L, Fang NF, Zhu HD (2012) Modeling the impacts of integrated small watershed management on soil erosion and sediment delivery: A case study in the Three Gorges Area, China. J Hydrol 438–439: 156–167.

28. Zhang Q, Lou Z (2011) The environmental changes and mitigation actions in the Three Gorges Reservoir region, China. Environ Sci Policy 14: 1132–1138.

29. Douglas-Mankin KR, Srinivasan R, Arnold JG (2010) Soil and Water Assessment Tool (SWAT) model: Current developments and applications.

30. T ASABE 53: 1423–1431.

31. USDA-SCS (1972) Hydrology Sect. 4, Soil Conservation Service National Engineering Handbook; Washington, DC.

32. Williams JR (1976) Flood routing with variable travel time or variable storage coefficients. T ASABE 12: 100–103.

33. Brown LC, Barnwell TO (1987) The Enhanced Stream Water Quality Models QUAL2E and QUAL2E-UNCAS: Documentation and User Manual; Athens.

34. Hong Q, Sun Z, Chen L, Liu R, Shen Z (2012) Small-scale watershed extended method for non-point source pollution estimation in part of the Three GorgesReservoir Region. Int J Environ Sci Tech 9: 595–604.

35. Abbaspour KC (2008) SWAT-CUP2: SWAT calibration and uncertainty programs - a user manual; Department of Systems Analysis, Integrated Assessment and Modelling (SIAM), Eawag, Swiss Federal Institute of Aquatic Science and Technology: Duebendorf.

36. Nash J, Sutcliffe J (1970) River forecasting using conceptual models, 1. A discussion of principles. J Hydrol 10: 282–290.

37. Shen ZY, Chen L, Chen T (2012) Analysis of parameter uncertainty in hydrological and sediment modeling using GLUE method: a case study of SWAT model applied to Three Gorges Reservoir Region, China. Hydrol Earth Syst Sci 16: 121–132.

38. Ma X, Li Y, Zhang M, Zheng F, Du S (2011) Assessment and analysis of nonpoint source nitrogen and phosphorus loads in the Three Gorges Reservoir Area of Hubei Province, China. Sci Total Environ 412–413: 154–161.

39. Li Q, Yu M, Lu G, Cai T, Bai X, et al. (2011) Impacts of the Gezhouba and Three Gorges reservoirs on the sediment regime in the Yangtze River, China. J Hydrol 403: 224–233.

40. Wu J, Cheng X, Xiao H, Wanga H, Yang LZ, et al. (2009) Agricultural landscape change in China's Yangtze Delta, 1942–2002: A case study. Agr Ecosyst Environ 129: 523–533.

41. Shen Z, Chen L, Liao Q, Liu R, Hong Q (2012) Impact of spatial rainfall variability on hydrology and nonpoint source pollution modeling. J Hydrol 472– 473: 205–215.

42. Shen Z, Hong Q, Yu H, Niu JF (2010) Parameter uncertainty analysis of nonpoint source pollution from different land use types. Sci Total Environ 408: 1971–1978.

43. Chen Q, Hooper DU, Lin S (2011) Shifts in Species Composition Constrain Restoration of Overgrazed Grassland Using Nitrogen Fertilization in Inner Mongolian Steppe, China. Plos One 6: e16909.

44. Ouyang W, Skidmore AK, Hao F, Wang T (2010) Soil erosion dynamics response to landscape pattern. Sci Total Environ 408: 1358–1366.

Chapter 8

Microbial Quality and Phylogenetic Diversity of Fresh Rainwater and Tropical Freshwater Reservoir

Rajni Kaushik[1,2], Rajasekhar Balasubramanian[1,2], and Hugh Dunstan[3]

[1]Singapore-Delft Water Alliance, National University of Singapore, Singapore, Singapore,

[2]Department of Civil and Environmental Engineering, National University of Singapore, Singapore, Singapore,

[3]School of Environmental and Life Sciences, the University of Newcastle, Callahan, NSW, Australia

ABSTRACT

The impact of rainwater on the microbial quality of a tropical freshwater reservoir through atmospheric wet deposition of microorganisms was studied for the first time. Reservoir water samples were collected at four different sampling points and rainwater samples were collected in the immediate vicinity of the reservoir sites for a period of four months (January to April, 2012) during the Northeast monsoon period. Microbial quality of all fresh rainwater and reservoir water samples was assessed based on the counts for the microbial indicators: Escherichia coli (E. coli), total coliforms, and Enterococci along with total heterotrophic plate counts (HPC). The taxonomic richness and phylogenetic relationship of the freshwater reservoir with those of the fresh rainwater were also assessed using 16 S rRNA gene clone library construction. The levels of E. coli were found to be in the range of 0 CFU/100 mL – 75 CFU/100 mL for the rainwater, and were 10–94 CFU/100 mL for the reservoir water. The sampling sites that were influenced by highway traffic emissions showed the maximum counts for all the bacterial indicators assessed. There was no significant increase in the bacterial abundances observed in the reservoir water immediately following rainfall. However, the composite fresh rainwater and reservoir water samples exhibited broad phylogenetic diversity, including sequences representing Betaproteobacteria, Alphaproteobacteria, Gammaproteobacteria, Actinobacteria, Lentisphaerae and Bacteriodetes. Members of the Betaproteobacteria group were the most dominant in both fresh rainwater and reservoir water, followed by Alphaproteobacteria, Sphingobacteria, Actinobacteria and Gammaproteobacteria.

INTRODUCTION

Airborne microorgansims can be transferred to aquatic systems through atmospheric fallout of coarse particles (dry deposition) [1] and rainfall (wet deposition) [2]–[3], leading to changes in the microbial composition of receiving water bodies [4]–[5]. The presence of bacterial pathogens in airborne particulate matter (PM) is of particular concern from the public health perspective as these aerosolized bacteria can form new cells in PM [6] and be metabolically active with the potential to biogeochemically mediate atmospheric chemistry [7]–[9]. The transfer of PM containing viable bacterial pathogens from the atmosphere to water bodies can affect not only ecotoxicology, but also human health through various exposure pathways.

Pathogenic episodes in lakes and reservoirs are often associated with rain events and riverine inflows [4], [10]. Excessive rainfall has been reported to be a significant contributor to historical waterborne disease outbreaks due to mobilization and transport of bacterial pathogens [10]–[12]. Stormwater runoff is a major cause of deterioration of surface water quality in urban areas. When rainfall occurs on paved surfaces, large volumes of water are swiftly carried to drains and discharged into receiving surface waters. Thus, the transport of microbial pollution from rain runoff to lakes and reservoirs is a major concern for management of natural waters worldwide [13]. Changes in the abundance of heterotrophic and coliform bacteria resident in stored water bodies have recently been reported in relation to incoming bacterial loads following rain events [5]. Hence, it is very important to understand the impact of rainfall carrying the live microbial aerosols on the quality of freshwater reservoirs in terms of changes in the abundance of pathogens, the microbial community composition and diversity.

Water treatment organizations and environmental protection agencies frequently use the presence of bacterial indicator organisms (e.g. fecal coliforms and enterococci) and their abundance in surface waters as a surrogate for the risk of contamination by actual pathogenic microorganisms [14]–[16]. Most of the studies reported in the literature characterized some bacteria residing in bulk water at various points in the drinking water supply system using cultivation-based approaches (14,17–18]. In general, heterotrophic plate counts (HPC) are used to assess the overall bacterial quality of drinking water, or natural waters [19]. Furthermore, most of the bacterial cells in natural communities are present in a viable but nonculturable (VBNC) state and therefore are non-culturable by current cultivation methods [20]–[21].

The real composition and dynamics of bacterial communities in freshwater resources remain largely unknown. In particular, limited studies have been conducted to determine the levels of bacterial pathogens in fresh rainwater prior to its collection and storage in order to assess its possible impact on the quality of roof-harvested rainwater and also that of aquatic systems upon deposition [1], [9]. The microbial quality of roof harvested rainwater is not only influenced by that of fresh rainwater, but also by the types of rooftop surfaces, the level of water tanks cleanliness, and the presence of insects' and birds' feces. However, no systematic study has been reported in the literature so far on the phylogenetic diversity of fresh rainwater describing the taxonomic richness of readily culturable organisms present in this water medium although such studies have been conducted for roof-harvested rainwater [22]. Therefore, there is a strong need for quantifying the levels of bacterial pathogens and phylogenetic diversity in both fresh rainwater and reservoir water following their simultaneous collection at the same sampling site of the reservoir.

Molecular approaches-based methods are good tools to elucidate composition of microbial communities residing in various aquatic environments [23]–[24]. These approaches exploit the use of rRNA as taxonomic marker for microorganisms [25]–[26]. The 16 S rRNA gene is tailor-made for microbial identification due to its universal presence in bacteria, extreme species sequence conservation and evolution-induced interspecies variability [27]. With the recent development and application of large-scale high throughput pyrosequencing-based method [28], community-wide spatial and temporal information on microbial community functional structure and potential activity can be rapidly obtained. Although the pyrosequencing-based approach is able to identify new sequences, it suffers from very high sensitivity to random sampling errors, dominant populations and contaminated non-target DNA [29]. The 16 S rRNA gene clone library method is a relatively old method that has been widely used in bacterial community analyses in various environments [30]–[32]. This method has been successfully applied to understand the role of freshwater bacterioplankton in global biogeochemical processes in the aquatic ecosystems [33]–[34]. However, the 16 S rRNA application to fresh rainwater has not been reported to date. The study of phylogenetic diversity in fresh rainwater and reservoir water could provide a basic understanding of the complex communities of environmental bacteria present in fresh rainwater and their impact on the composition of microbial communities in reservoir water intended for human consumption. The objectives of this work were to (1) study the impact of microbial loading of fresh rainwater on the water quality of a tropical reservoir in Singapore (characterized by heavy rainfall) upon deposition using total bacteria and traditional bacterial indicators and (2) examine the kind of bacterial diversity that fresh rainwater and reservoir water harbor using traditional 16 s rRNA cloning and sequencing method.

MATERIALS AND METHODS

Sampling

Singapore's climate is characterized by uniform temperature and pressure, high humidity, and abundant rainfall. Singapore receives an annual rainfall of about 2400 mm. There are no distinct wet, or dry seasons as rainfall occurs every month of the year. The two main seasons, based on the prevailing dominant winds, are the Northeast monsoon season (from late November to March), and the Southwest monsoon season (from late May to September). April to early May and October to early November are generally the transitional months separating the monsoons. December is usually the wettest month with an average rainfall of 280 mm.

Both rainwater samples and reservoir samples were collected at a tropical reservoir (coordinates-1°22'03"N 103°48'07"E) as part of the Singapore-Delft Water Alliance project. There was no need for any special permission to carry out the water sampling, and the activities did not involve endangered or protected species. Reservoir water samples were collected from four different sites at the reservoir using a grab sampling method along with rainwater samples. Both types of water samples were collected concurrently at the same four sites (No. 1, 2, 3 and 4). The reservoir has a capacity of 27.8 million of water over 304 hectares of water surface. Site 1 was located at the centre of the reservoir with no land use. Site 2 was located at the corner of the reservoir with influences from a local drainage system. Site 3 was situated near a major highway and site 4 was near a golf course, having anthropogenic influence from this land use type (Fig.1).

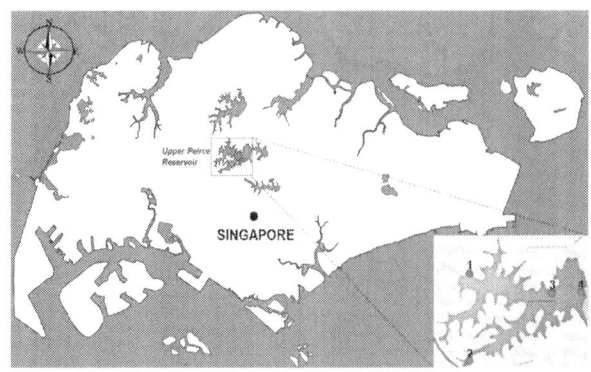

Figure 1: Outline of the four sampling sites at the tropical reservoir of Singapore used in this study.

Both rainwater and reservoir water samples were collected from the sampling locations at regular weekly intervals and within 24 h of rainfall events between January 2012 and April 2012. A total of 33 rain samples on an event-to-event basis were collected (n = 6, 5, 12, and 10 for January, February, March and April, respectively). On each sampling occasion two replicate grab samples (1 L) were collected 1 m from the shore at a depth of ~0.5 m in sterile containers. 1 Liter of fresh rainwater samples was collected in sterilized glass bottles (amber colored glass bottles with a sterilized funnel attached to them) placed at all the four sites used for reservoir sampling. Both fresh rainwater and reservoir water samples (immediately after rainfall) collected from all the four sites were transferred to 2 L sterilized bottles for microbiological and chemical analyses. These water samples were transported to the laboratory in a chilled-cold box and processed within 6 h of collection.

Microbiological Analyses

Concentrations of total coliform bacteria and E. coli were determined using the m-ColiBlue24 membrane filtration system (Millipore, Cat #M00PMCB24, Bedford, Massachusetts). For all the collected water samples, 100 mL of the sample was filtered onto cellulose ester membranes using vacuum filtration and the membrane was then incubated for 24 h in sterile petri dishes containing absorbent pads soaked with 2 mL of m-ColiBlue24 broth at 37°C. This was performed in duplicate for all the samples. The colonies in blue color were indicative of E. coli, while total coliforms were enumerated by counting the colonies in red color. The average CFU/100 mL values obtained for rain events in each month for E. coli and total coliforms were estimated.

HPC and Enterococci counts were also determined based on average counts obtained for all the samples collected at the four sites in this study. Briefly, for HPC enumeration, one mL of each water sample was serially diluted and the dilutions were aseptically plated in duplicates onto a plate count agar (Sigma-Aldrich, USA) and incubated at 37°C for a maximum duration of 48 h. The average colony counts were expressed as CFU/mL.

For Enterococci, the enumeration was performed as per the USEPA Method [35]. In brief, 100 mL of the water sample was filtered onto cellulose esters membranes using vacuum filtration in duplicate and the membranes were then placed on top of the membrane-Enterococcus Indoxyl-β-D-Glucoside Agar (mEI Agar, BD, NJ, USA) incubated for 24 h at 41±0.5°C. Colonies with a blue halo, regardless of color, were enumerated as Enterococci. The colony counts were expressed as average CFU/100 mL.

DNA Extraction

One liter of each fresh rainwater and reservoir water samples (duplicates for each sampling site and date) was first filtered through a pre-combusted glass fiber filter (90-mm diameter, GF/F, Whatman) followed by a 0.22 μm hydrophilic polycarbonate filter (Millipore Corporation, Bedford, MA). The filters were cut into small pieces and suspended in 1.0 mL of phosphate buffered saline and DNA was extracted using Ultraclean Microbial DNA Isolation kits (MO BIO laboratories, Carlsbad, CA) according to the manufacturer›s instructions. The extracted DNA solution was stored at −80°C.

Construction of 16 S rRNA Gene Libraries

Clone libraries were constructed from composite DNA samples of fresh rainwater and composite DNA samples of reservoir water obtained at the four sites. Briefly, the 16 S rRNA gene was amplified from pooled environmental DNA samples using primers 27f (5'-AGAGTTTGATCMTGGCTCAG-3') and 1492r (5'-TACCTTGTTACGACTT-3') [36]. The PCR products were purified using UltraClean PCR Clean-up DNA purification kits (MO BIO Laboratories) and made up to 30 µL with water. Amplified fragments were cloned into TOPO-TA plasmids using the TOPO-TA cloning kit (Invitrogen, Carlsbad, CA) and transferred intoEscherichia coli DH5a cells (TakaraBio, Otsu, Japan) to construct 16 S rRNA gene libraries.

Cloned plasmid inserts were amplified directly from cells as described [37] using vector primers. The 16 S rRNA gene portion of the cloned DNA was initially sequenced using the ABI Prism BigDye terminator v3.1 and cycle sequencing kit (PE Applied Biosystems). We sequenced 150 clones for each bacterial gene library.

Phylogenetic Analyses

Cloned gene sequences were vector-trimmed and aligned using the NAST (Nearest Alignment Space Termination) algorithm for creating multiple sequence alignments [38]. NAST aligned sequences were chimera checked with chimera slayer. All bioinformatics processes were performed in the MOTHUR environment [39]. Classification of sequences was done using Bayesian classifier method implemented in MOTHUR against Greengenes database [40]. The division level groupings were determined by taxonomic assignment performed by the Ribosomal Database Project 10.0 Classifier tool [41]. Pairwise distances of the aligned clone sequences were calculated and Operational Taxonomic units (OTUs) were grouped using the average neighbor method with a cut-off of 0.03 sequence similarity. Coverage of clone libraries was calculated according to the method recommended by Good (1953) [42]. A Phylogenetic tree was constructed using the neighbor- joining method of ARB with Jukes-Cantor correction model and with 1000 bootstrap replications [43].

Community Statistical Analyses

Microbial quality data were subjected to the student t-test for determining statistical significance. For microbial diversity, analyses of beta diversity with replicate data from rainwater and its respective reservoir water were used for comparison of the two microbial communities. All profiles were inter-compared in a pair-wise fashion to determine a dissimilarity score and stored in a distance matrix. The Unifrac distance metric, as described in Lozupone et al. 2006, was utilized for the phylogenetic distance between OTUs to determine the dissimilarity between the two communities [44]. Weighted Unifrac was used considering the OTU abundance along with the Adonis test of statistical significance ($p<0.05$), which is a non-parametric multivariate analysis of variance (MANOVA) with the Adonis function and utilizes the sample-to-sample distance matrix directly for finding significant differences among these communities [45].

RESULTS AND DISCUSSION

Microbiological Quality of Fresh Rainwater and Reservoir Water

Microbial quality is usually assessed by measuring 'fecal indicator bacteria' (also referred to as fecal indicator organisms, or FIOs) that are generally opportunistic pathogens, present in large numbers in fecal materials. Their presence in water samples is used to indicate the presence of fecal pollution and the possibility that fecal associated pathogens may also be present. The most commonly examined FIOs are thermotolerant coliforms (also termed fecal coliforms),Escherichia coli and intestinal Enterococci (also termed fecal streptococci). E. coli are considered the best indicators of fecal contamination in water. The presence of thermotolerant coliforms/E. coli in water is unacceptable from the public health perspective as it indicates that a major health risk exists. However, it should be noted that certain members of the coliform group live outside of the gastrointestinal tract in the environment and may create a false indication of fecal contamination [46]. Also, several strains of Escherichia coli which are of environmental origin can be thermotolerant, thus giving false positive results [47]. It was therefore necessary to evaluate a range of relevant indicator organisms and total bacteria.

Figs. 2, 3 and 4 show these microbial indicators for both fresh rainwater and reservoir water over the four months of sampling. Levels of thermotolerant coliforms/E. coli are expressed as colony forming units (CFU) per 100 mL (CFU/100 mL). The levels of E. coli were found to be in the range of 0 CFU/100 mL–75 CFU/100 mL for fresh rainwater and in the range of 10 CFU/100 mL–94 CFU/100 mL in reservoir water (Fig. 2). According to the WHO guidelines for treated drinking water (2006), the levels of E. coli should be less than 1 CFU/100 mL [48]. Thus, fresh rainwater is non potable for direct consumption and needs to be treated for potable purposes.

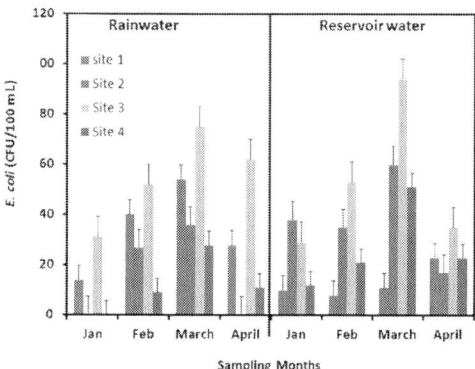

Figure 2: E. coli counts from both rainwater samples and reservoir samples collected from four sites from January to April, 2012.

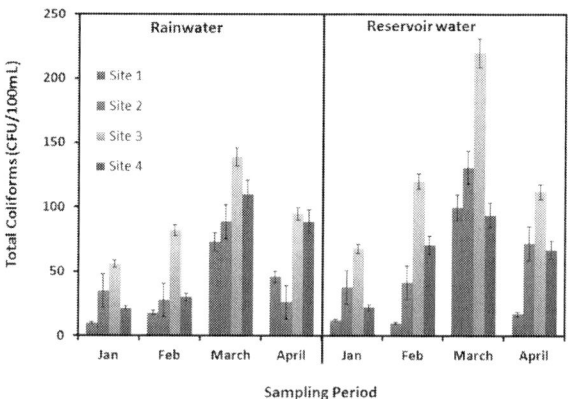

Figure 3: Total coliforms counts from both rainwater samples and reservoir samples collected at four sites from January to April, 2012.

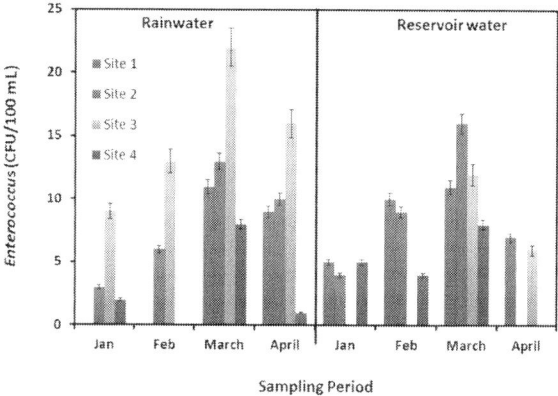

Figure 4: Enterococcus counts from both rainwater samples and reservoir samples collected at four sites from January to April, 2012.

Total coliforms were previously considered indicators of fecal contamination. The NHMRC and AWRC "Guidelines for Drinking Water Quality in Australia" (2004) do not consider total coliforms as useful indicators of fecal contamination in the absence of E. coli, and have not proposed any guideline value for total coliforms [49]. As the case with HPC, total coliforms can also be used as an indicator for the effectiveness of any treatment program. However, it should be noted that while fecal coliforms are enteric organisms and therefore virtually exclusively of fecal origin, the total coliform count may comprise environmental organisms that are not necessarily common to the digestive tracts of vertebrate animals [50]. The total coliforms were found in the range of 10 CFU/100 mL–220 CFU/100 mL in reservoir water as compared to 10 CFU/100 mL–139 CFU/100 mL in fresh rainwater (Fig. 3). The highest counts were obtained from site #3 in both rainwater and reservoir water, which is near a major highway with residential buildings in the vicinity. This anthropogenic influence could be due to the abundance of airborne microbial pathogens in urban areas with high population density [1], [9]; Turkum et al. (2008) reported that most of the species found in rainwater were derived from aerosol and gas-phase components [51].

Enterococci are a specific group of bacteria that are found in high numbers in both human and animal faeces, and are therefore a valuable indicator for determining the extent of fecal contamination of a water source [16], [52]–[53]. Enterococci were found to be in the range of 0

CFU/100 mL–11 CFU/100 mL in fresh rainwater and 0 CFU/100 mL–35 CFU/100 mL in the reservoir water (Fig. 4). Overall, the four microbial indicators were found to be the highest at site #3 in both types of water samples. Thus, the results of our microbiological indicator analyses suggested that both the fresh rainwater and the reservoir water samples are not suitable for human consumption without any treatment similar to the findings on the roof harvested rainwater [22], [52], [54].

Total HPC bacteria of the reservoir water samples were two orders of magnitude higher than those of the rainwater. The baseline levels for the reservoir water ranged from 330 CFU/mL to 7.9×10^4 CFU/mL, as compared to 280 CFU/mL to 7.2×10^2 CFU/mL in fresh rainwater (Fig. 5). The reason for the reservoir water having two fold higher magnitudes of total HPC bacteria and slightly higher number of total coliforms could be the additional contribution from sediments apart from rainfall and riverine inflows. HPC bacteria are carried by runoff into the reservoir from highland agricultural areas during intensive rainfall events. Thus, rain events may lead to an inflow of high nutrient concentrations as well as high loads of microbes [55]–[56]. Based on the microbial indicator numbers, there was no statistically significant difference between the rainwater and reservoir water quality with respect to E. coli and Enterococci. However, the microbial quality of rainwater and reservoir was found to be significantly different ($p \leq 0.05$) for total heterophic counts and total coliforms according to the t-test. It was also observed that the concentrations of all the microbial indicators were higher in the month of March as compared to those in other months. This can be attributed to a higher frequency of rainfall in March as compared to the relatively dry period in the months of January and February.

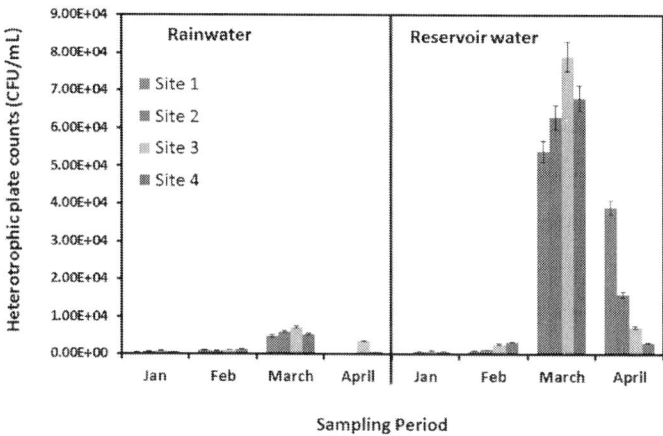

Figure 5: Total heterotrophic plate count bacteria from both rainwater samples and reservoir samples collected at four sites from January to April, 2012.

It should be noted that E. coli and Enterococci are indicative of recent fecal contamination as compared to heterophic counts and total coliforms. The influence of E. coli and Enterococci of atmospheric origin on the reservoir water quality appears to be relatively less as compared to that of total bacteria HPC and total coliforms, this could due to their lower abundance in the atmosphere and/or the influence of physical processes such as advection, dispersion, diffusion resulting in low residence time [22]. Other processes such as competitive exclusion and nutrient change may also play role to regulate the survival of these incoming bacteria [5].

Microbial Composition and Phylogenetic Analyses of Fresh Rainwater and Reservoir Water

Microbial communities are fundamental to the functioning of aquatic ecosystems. With the availability of metagenomic technologies and approaches [57], it is important to know the changes in total microbial diversity of both rainwater and reservoir water in addition to the changes in microbial indicator numbers. Gaining insights into the total microbial community would provide a better understanding of its role in mediating aquatic ecological processes and biogeochemical cycling [57]–[58]. However, the roles of most microorganisms in natural systems are unclear as most of them cannot be cultivated for investigation by current conventional culturing methods [59]. The introduction of culture-independent molecular methods has shed light on the determination of community compositions, and laid the foundation for gaining a deep understanding of aquatic microbial ecology. Rainwater has received considerable attention as a potential alternative source of potable and non-potable water in regions where there is water scarcity [60]. Rainwater collection is an ancient practice and currently still being practiced, especially in areas with no running water. However, the use of 16 S rRNA sequencing techniques has not been applied to fresh rainwater investigations. Thus, there is a lack of information on its microbial communities and their role in aquatic microbial ecology.

In this study, 10 classes of bacteria were detected in fresh rainwater and four classes of bacteria in reservoir water. In fresh rainwater, sequences were affiliated with Betaproteobacteria, Alphaproteobacteria, Sphingobacteria, Actinobacteria, Gammaproteobacteria, Lentisphaerae, CH21, Phycisphaerae, Chlorobia and Spirochaetes. In contrast, the reservoir water library detected sequences affiliated with only Betaproteobacteria, Alphaproteobacteria, Sphingobacteria and Gammaproteobacteria. Betaproteobacteria was found to be the dominant class in both the libraries (Fig. 6). The fresh rainwater had higher diversity and taxonomic richness at the class level than those of reservoir water as the numbers of OTUs were found to be 63 and 29 for fresh rainwater and reservoir water, respectively. However, the proportion of Alphaproteobacteria in the bacterial community was found to be higher in the reservoir water than in the fresh rainwater. Alphaproteobacteria, at least at the class level, are resistant to predation: their relative higher abundance in reservoir water may be due to the capacity to degrade recalcitrant organic compounds such as humic substances accompanied by a tendency of members of the Alphaproteobacteria to form filaments, aggregates, or Caulobacter-like stalked cells that sometimes can make up a majority of the Alphaproteobacteria population in freshwater lakes [61].

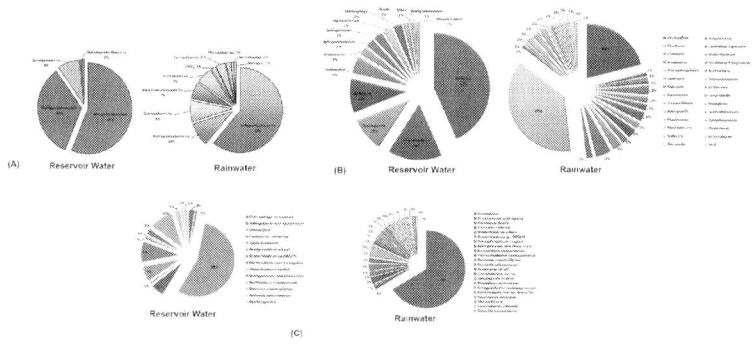

Figure 6: Bacteria taxon richness on (A) class level (B) genus level and (C) species level observed in the fresh rainwater and reservoir water.

The taxonomic richness observed in both types of water samples and their relative distribution appears to be similar to that observed in many freshwater and marine systems, as determined by both cultivation and molecular techniques [22], [62]–[63]; these findings are consistent with the results from previous studies where the dominance of Betaproteobacteria has been observed consistently in freshwater systems, particularly among free-living groups, while   and   sub-classes appear to dominate among particle-attached groups [64]. The high abundance and amenability to culturing have contributed to the Betaproteobacteria being the best-studied group in freshwater lakes [61]. Venter et al. (2004) carried out a comprehensive genomic study of natural waters at an oligotrophic ocean site in the Sargasso Sea, and found an abundant bacterial distribution dominated by Proteobacteria of α, β and γ sub-classes, followed by Actinobacteria and Bacteroidetes [65]. These findings are similar to those from the fresh rainwater microbial community analyses in the current study as well as from the study conducted by Evans et al., 2009 in rainwater tanks [22]. The class Betaproteobacteria was the dominant and the most diverse group in freshwater mesocosm and this dominance is suggested to be associated with their ability to respond quickly to nutrient additions [66]–[67]. The predominance of Proteobacteria in fresh rainwater can also be suggestive of the generally clean oligotrophic nature of fresh rainwater.

Actinobacteria are Gram-positive bacteria representing the most abundant group (50–70%) of total bacteria in various freshwater habitats [68]–[69] were also found to be present in fresh rainwater. Previous studies have found that members of Actinobacteria contributed to glucose assimilation and heterotrophic nitrification [70]–[71] and played a key role in nutrient and energy cycling in aquatic habits [71]. Also most interestingly, the UV stress resistance has also been postulated to be one of the reasons for the success of the Actinobacteria in the upper waters of lakes, which often have high UV transparency. Warnecke and colleagues showed a significant positive relationship between Actinobacteria abundance and UV transparency in study on mountain lakes [68]. In addition to UV protection, many members of the Actinobacteria are capable of producing spores, allowing them to survive long periods of desiccation. Thus, strong UV protection and desiccation resistance via spore formation, together with the known small cell size of these organisms, would make the freshwater Actinobacteria particularly suitable for aerial dispersal, thereby explaining their ubiquitous representation in globally dispersed lakes. However, to date, these taxa have rarely been identified in any air samples [61]. This is the first study to-date to detect their presence in fresh rainwater suggesting their ability to survive in airborne particles that tend to be scavenged by rainwater during rain events [1]. This wet-deposition process enables the global distribution of the microorganisms of atmospheric origin in aquatic ecosystems [66].

Phylogenetic tree-based microbiome comparison of the fresh rainwater and reservoir water samples was performed by the neighbor-joining method of experimentally observed OTUs and reference type strains with each leaf representing an OTU displays counts of sequences observed in each sample (Fig. 7). Notable differences in the frequencies of the some of the bacterial genera were observed. Bradyrhizobium, Chitinophaga, Caulobacter and Ralstoniawere found to be more abundant in reservoir water than in fresh rainwater whereas Curvibacterwas found to be more abundant in fresh rainwater than in reservoir water. However, unlike taxonomic richness, the coverage of the fresh rainwater clone library (150 sequences from pooled DNA) as calculated based on the species OTU (97% 16 S rRNA gene sequence identity) by Good›s Clone Coverage [42] was low (77%) as compared to the coverage of the reservoir water clone library (92%).

Figure 7: Taxonomic tree indicating the phylogenetic distribution of all bacterial groups in both rainwater and reservoir water samples.

The OTU with the greatest difference in relative abundance was associated with Curvibacter for fresh rainwater and with Ralstonia for reservoir water. Curvibacter and Ralstonia belong to theBurkholderiales order of Betaproteobacteria, under the taxa Comamonadaceae that is reported to be the most abundant typical freshwater fast growing and nutrient-loving group [31], [72]. Rainwater and reservoir samples formed separate groups in ordination analysis, but were not found to be significantly dissimilar (Adonis p-value = 0.125) using OTU abundance metrics. The same diversity of bacteria being present in two different environmental samples suggests that the microbiological composition of the atmosphere and that of the hydrosphere are intimately linked with each other through wet and dry deposition mechanisms. The type of bacterial diversity present in fresh rainwater, the general abundance distribution, and the resemblance of the composition to that of reservoir, have indicated the likely existence of definable micro ecosystems in fresh rainwater and their impact on other aquatic systems. The functional operation of a stable micro-ecology, dominated by well-adapted core resident groups, may have implications with regard to the harvesting of rainwater and for management of water resources.

CONCLUSIONS

This study focused on the investigation of the microbial diversity of fresh rainwater and that of reservoir water with a concurrent assessment of bacterial counts and species profiles in a tropical region, characterized by abundant rainfall throughout the year, for the first time. The results of the study suggested that despite the presence of bacterial pathogens in rainwater and their wet deposition through rainfall, there was no significant increase in the microbial loading of the tropical reservoir water. It appears that the incoming bacterial loads from the rainwater entering the reservoir either undergo dilution, or are not sustained for a long period of time due to the influence of physical processes such as advection, dispersion and diffusion and other processes (competitive exclusion and nutrient change) that may act to regulate the survival of incoming bacteria. Thus, despite the presence of microbial pathogens, rainwater harvesting in large water catchment areas is a promising freshwater resource following treatment. These findings on the composition of aquatic microbial communities in the reservoir water and fresh rainwater by 16 s rRNA clone libraries indicated that the members of Betaproteobacteriadominated both the communities. The OTU with the greatest difference in the relative abundance for fresh rainwater was Curvibacter whereas for reservoir water it was Ralstonia. Actinobacter was detected in fresh rainwater, suggesting their presence in airborne particles as well. The fresh rainwater showed greater taxonomic richness than that of the reservoir water.

However, due to lower coverage in traditional cloning method, our understanding about the bacterial composition and phylogenetic diversity of the bacterial community in the freshwater is still limited. Hence, high-throughput molecular tools such as pyrosequencing and hybridization to PhyloChip array that are able to elucidate not only cultivable bacteria, but also viable-but-not-cultivable (VBNC) may extend and expand our understanding of the tropical freshwater reservoir microbial communities. More in-depth studies on the linkage between atmospheric inputs of microorganisms and the microbial diversity of freshwater resources are warranted. This research group has recently performed study of reservoir microbial diversity using PhyloChip microarrays at the same reservoir in the presence and absence of rain events (work in progress), which may add more insights into differences between the traditional cloning and metagenomics of freshwater bacterial diversity.

ACKNOWLEDGMENTS

The authors gratefully acknowledge the support and contributions to Singapore-Delft Water Alliance, National University of Singapore (NUS) for the support provided to the pursuit of this study. Rajni Kaushik thanks SDWA for supporting her Ph.D. study. The technical assistance for sampling provided by Chen Sijing is gratefully acknowledged.

AUTHOR CONTRIBUTIONS

Conceived and designed the experiments: RK RB. Performed the experiments: RK. Analyzed the data: RK RB HD. Contributed reagents/materials/analysis tools: RK RB. Contributed to the writing of the manuscript: RK RB HD.

REFERENCES

1.   Kaushik R, Balasubramanian R (2012) Assessment of bacterial pathogens in fresh rainwater and airborne particulate matter using Real-Time PCR. Atmos Environ 46: 131–139. doi: 10.1016/j.atmosenv.2011.10.013

2.   Womack AM, Bohannan BJM, Green JL (2010) Biodiversity and biogeography of the atmosphere. Philos Trans R Soc Lond B Biol Sci 365: 3645–3653. doi: 10.1098/rstb.2010.0283

3.   Kaushik R, Balasubramanian R, de la Cruz AA (2012) Influence of air quality on the composition of microbial pathogens in fresh rainwater. Appl Environ Microbiol 78: 2813–2818. doi: 10.1128/aem.07695-11

4.   Brookes JD, Antenucci J, Hipsey M, Burch MD, Ashbolt NJ, et al. (2004) Fate and transport of pathogens in lakes and reservoirs. Environ Int 30: 741–759. doi: 10.1016/j.envint.2003.11.006

5.   Martin AR, Coombes PJ, Harrison TL, Dunstan RH (2010) Changes in abundance of heterotrophic and coliform bacteria resident in stored water bodies in relation to incoming bacterial loads following rain events. J Environ Monitor 12: 255–260. doi: 10.1039/b904042k

6.   Dimmick RL, Wolochow H, Chatigny MA (1979) Evidence that bacteria can form new cells in airborne particles. Appl. Environ. Microbiol 37: 924–927.

7.   Ariya P, Nepotchatykh O, Ignatova O, Amyot M (2002) Microbiological degradation of atmospheric organic compounds. Geophys Res Lett 29: 1–4. doi: 10.1029/2002gl015637

8.   Amato P, Ménager M, Sancelme M, Laj P, Mailhot G, et al. (2005) Microbial population in cloud water at the Puy de Dôme: Implications for the chemistry of clouds. Atmos Environ 39: 4143–4153. doi: 10.1016/j.atmosenv.2005.04.002

9.   Kaushik R, Balasubramanian R (2013) Discrimination of viable from nonviable Gram-negative bacterial pathogens in airborne particles using propidium monoazide-assisted qPCR. Sci Total Environ 449: 237–243. doi: 10.1016/j.scitotenv.2013.01.065

10.   Auld H, MacIver D, Klaassen J (2004) Heavy rainfall and waterborne disease outbreaks: The Walkerton example. J Toxicol Environ Health A 67: 1879–1887. doi: 10.1080/15287390490493475

11.   Curriero FC, Patz JA, Rose JB, Lele S (2001) The association between extreme precipitation and waterborne disease outbreaks in the United States, 1948–1994. Am J Public Health 91: 1194–1199. doi: 10.2105/ajph.91.8.1194

12.   Ferguson C, Husman AMD, Altavilla N, Deere D, Ashbolt N (2003) Fate and transport of surface water pathogens in watersheds. Crit

Rev Env Sci 33: 299–361. doi: 10.1080/10643380390814497

13. Olson BH, Nagy LA (1984) Microbiology of potable water. Adv Appl Microbiol 30: 73–132. doi: 10.1016/s0065-2164(08)70053-4

14. Ward NR, Wolfe RL, Justice CA, Olson BH (1986) The identification of gram-negative, non-fermentative bacteria from water: problems and alternative approaches to identification. Adv Appl Microbiol 31: 293–365. doi: 10.1016/s0065-2164(08)70446-5

15. Edberg SC, Rice EW, Karlin RJ, Allen MJ (2000) Escherichia coli: the best biological drinking water indicator for public health protection. J Appl Microbiol 88: 106S–116S. doi: 10.1111/j.1365-2672.2000.tb05338.x

16. Sidhu J, Ahmed W, Toze S (2013) Application of Microbial Source Tracking Toolbox to Identify Sewage Contamination in Stormwater Run-off in Brisbane AWA Water. 40: 81–85.

17. LeChevallier MW, Babcock TM, Lee RG (1987) Examination and characterization of distribution system biofilms. Appl Environ Microbiol 53: 2714–2724.

18. Sartory DP (2004) Heterotrophic plate count monitoring of treated drinking water in the UK: a useful operational tool. Int J Food Microbiol 92: 297–306. doi: 10.1016/j.ijfoodmicro.2003.08.006

19. Oliver JD (2000) The public health significance of viable but nonculturable bacteria. In: R. RColwell and D. JGrimes (ed). Nonculturable microorganisms in the environment. ASM Press, Washington D.C., pp. 277–300.

20. Giovannoni SJ, Britschgi TB, Moyer CL, Field KG (1990) Genetic diversity in Sargasso Sea bacterioplankton. Nature 345: 60–63. doi: 10.1038/345060a0

21. Szewzyk U, Szewzyk R, Manz W, Schleifer KH (2000) Microbiological safety of drinking water. Annu Rev Microbiol 54: 81–127. doi: 10.1146/annurev.micro.54.1.81

22. Evans CA, Coombes PJ, Dunstan RH, Harrison T (2009) Extensive bacterial diversity indicates the potential operation of a dynamic micro-ecology within domestic rainwater storage systems. Sci Total Environ 407: 5206–5215. doi: 10.1016/j.scitotenv.2009.06.009

23. Woese CR (1987) Bacterial evolution. Microbiol Rev 51: 221–271.

24. Höfle MG, Haas H, Dominik K (1999) Seasonal dynamics of bacterioplankton community structure in a eutrophic lake as determined by 5S rRNA analysis. Appl Environ Microbiol 65: 3164–3174.

25. Altmann D, Stief P, Amann R, De Beer D, Chramm A (2003) In situ distribution and activity of nitrifying bacteria in freshwater sediment. Environ Microbiol 5: 798–803. doi: 10.1046/j.1469-2920.2003.00469.x

26. Tringe SG, Hugenholtz P (2008) A renaissance for the pioneering 16S rRNA gene. Curr Opin Microbiol 11: 442–446. doi: 10.1016/j.mib.2008.09.011

27. Schwarzenbach RP, Egli T, Hofstetter TB, von Gunten U, Wehrli B (2010) Global Water pollution and human health. Annu Rev Env Resour 35: 109–136. doi: 10.1146/annurev-environ-100809-125342

28. Huber JA, Mark Welch D, Morrison HG, Huse SM, Neal PR, et al. (2007) Microbial population structures in the deep marine biosphere. Science 318: 97–100. doi: 10.1126/science.1146689

29. Zhou J, Kang S, Schadt CW, Garten CT Jr (2008) Spatial scaling of functional gene diversity across various microbial taxa. Proc Natl Acad Sci USA 105: 7768–7773. doi: 10.1073/pnas.0709016105

30. Hiorns WD, Methe BA, NierzwickiBauer SA, Zehr JP (1997) Bacterial diversity in Adirondack Mountain lakes as revealed by 16 S rRNA gene sequences. Appl Environ Microbiol 63: 2957–2960.

31. Zwart G, Hiorns WD, Methe B, van Agterveld MP, Huismans R, et al. (1998) Nearly identical 16 S rRNA sequences recovered from lakes in North America and Europe indicate the existence of clades of globally distributed freshwater bacteria. Syst Appl Microbiol 21: 546–556. doi: 10.1016/s0723-2020(98)80067-2

32. Wu X, Xi WY, Ye WJ, Yang H (2007) Bacterial community composition of a shallow hypertrophic freshwater lake in China, revealed by 16S rRNA gene sequences. FEMS Microbiol Ecol 61: 85–96. doi: 10.1111/j.1574-6941.2007.00326.x

33. Zehr JP, Ward BB (2002) Nitrogen cycling in the ocean: new perspectives on processes and paradigms. Appl Environ Microbiol

68: 1015–1024. doi: 10.1128/aem.68.3.1015-1024.2002

34. Mason OU, Di Meo-Savoie CA, Van Nostrand JD, Zhou JZ, Fisk MR, et al. (2009) Prokaryotic diversity, distribution, and insights into their role in biogeochemical cycling in marine basalts. ISME J 3: 231–242. doi: 10.1038/ismej.2008.92

35. U.S. Environmental Protection Agency (USEPA) (2002) Method 1600: Enterococci in water by membrane filtration using membrane-Enterococcus Indoxyl- -D-Glucoside Agar (mEI). EPA-821-R-02-022. Office of Water, Washington, D.C.

36. Weisburg WG, Barns SM, Pelletier DA, Lane DJ (1991) 16S ribosomal DNA amplification for phylogenetic study. J Bacteriol 173: 697–703.

37. Vergin KL, Rappe' MS, Giovannoni SJ (2001) Streamlined method to analyze 16 S rRNA gene clone libraries. Biotechniques 30: 938–940.

38. DeSantis TZ, Hugenholtz P, Keller K, Brodie EL, Larsen N, et al. (2006) NAST: a multiple sequence alignment server for comparative analysis of 16S rRNA genes. Nucleic Acids Res 34: W394–W399. doi: 10.1093/nar/gkl244

39. Schloss PD, Westcott SL, Ryabin T, Hall JR, Hartmann M, et al. (2009) Introducing mothur: Open-Source, Platform-Independent, Community-Supported Software for Describing and Comparing Microbial Communities. Appl Environ Microbiol 75: 7537–7541. doi: 10.1128/aem.01541-09

40. McDonald D, Price MN, Goodrich J, Nawrocki EP, DeSantis TZ, et al. (2012) An improved Greengenes taxonomy with explicit ranks for ecological and evolutionary analyses of bacteria and archaea. ISME J 6: 610–618. doi: 10.1038/ismej.2011.139

41. Cole JR, Wang Q, Cardenas E (2009) The Ribosomal Database Project: improved alignments and new tools for rRNA analysis. Nucleic Acids Res 37: D141–D145. doi: 10.1093/nar/gkn879

42. Good IJ (1953) The Population Frequencies Of Species And The Estimation Of Population Parameters. Biometrika 40: 237–264. doi: 10.1093/biomet/40.3-4.237

43. Ludwig W, Strunk O, Westram R, Richter L, Meier H, et al. (2004) ARB: a software environment for sequence data. Nucleic Acids Res 32: 1363–1371. doi: 10.1093/nar/gkh293

44. Lozupone C, Hamady M, Knight R (2006) UniFrac—an online tool

for comparing microbial community diversity in a phylogenetic context. BMC Bioinformatics 7: 371–385.

45. He Z, Xu M, Deng Y, Kang S, Kellogg L, et al. (2010) Metagenomic analysis reveals a marked divergence in the structure of belowground microbial communities at elevated CO_2. Ecol Lett 13: 564–575. doi: 10.1111/j.1461-0248.2010.01453.x

46. Pisciotta JM, Rath DF, Stanek PA, Flanery DM, Harwood VJ (2002) Marine bacteria cause false-positive results in the Colilert-18 rapid identification test for Escherichia coli in Florida waters. Appl Environ Microbiol 68: 539–544. doi: 10.1128/aem.68.2.539-544.2002

47. Edberg SC, Allen MJ, Smith DB, Kriz NJ (1990) Enumeration of total coliforms and Escherichia coli from source water by the defined substrate technology. Appl Environ Microbiol 56: 366–369.

48. WHO (2006) Guidelines for drinking-water quality. Geneva: World Health Organization.

49. National Health and Medical Research Council (2004) Guidelines for drinking water quality in Australia. National Health and Medical Research Council (NHMRC & AWRC), Canberra, Australia.

50. Evans CA, Coombes PJ, Dunstan RH (2006) Wind, rain and bacteria: The effect of weather on the microbial composition of roof-harvested rainwater. Water Res 40: 37–44. doi: 10.1016/j.watres.2005.10.034

51. Turkum A, Pekey H, Pekey B, Tuncel G (2008) Investigating relationships between aerosol and rainwater compositions at different locations in Turkey. Atmos Res 89: 315–323. doi: 10.1016/j.atmosres.2008.03.010

52. Ahmed W, Richardson K, Sidhu J, Toze S (2012) Escherichia coli and Enterococcus spp. in rainwater tank samples: comparison of culture-based methods and 23 S rRNA gene quantitative PCR assays. Environ Sci Technol 46: 11370–11376. doi: 10.1021/es302222b

53. Savichtcheva O, Okabe S (2006) Alternative indicators of fecal pollution: Relations with pathogens and conventional indicators, current methodologies for direct pathogen monitoring and future application perspectives. Water Res 40: 2463–2476. doi:

10.1016/j.watres.2006.04.040

54. Ahmed W, Sidhu J, Toze S (2013) Faecal Indicators and pathogens in Potable Rainwater tanks in Southeast Queensland. AWA Water 40: 88–92.

55. Merz J, Dangol PM, Dhakal MP, Dongol BS, Nakarmi G, et al. (2006) Rainfall–runoff events in a middle mountain catchment of Nepal. J Hydrol 331: 446–58. doi: 10.1016/j.jhydrol.2006.05.030

56. .Sargaonkar A (2006) Estimation of land use specific runoff and pollutant concentration for Tapi River Basin in India. Environ Monit Assess 117: 491–503. doi: 10.1007/s10661-006-0769-2

57. Edwards RA, Rohwer F (2005) Viral metagenomics. Nat Rev Microbiol 3: 504–510. doi: 10.1038/nrmicro1163

58. Tseng CH, Chiang PW, Shiah FK, Chen YL, Liou JR, et al. (2013) Microbial and viral metagenomes of a subtropical freshwater reservoir subject to climatic disturbances. ISME J 7(12): 2374–86. doi: 10.1038/ismej.2013.118

59. Amann RI, Ludwig W, Schleifer KH (1995) Phylogenetic identification and in situ detection of individual microbial cells without cultivation. Microbiol Rev 59: 143–169.

60. Meera V, Mansoor Ahammed M (2006) Water quality of rooftop rainwater harvesting systems: a review. J. Water Supply Res Technol.-AQUA 55 (8) 257–267. doi: 10.2166/aqua.2006.052

61. Newton RJ, Jones SE, Eiler A, McMahon KD, Bertilsson S (2011) A guide to the natural history of freshwater lake bacteria. Microbiol Mol Biol Rev 75: 14–49. doi: 10.1128/mmbr.00028-10

62. Newton RJ, Kent AD, Triplett EW, McMahon KD (2006) Microbial community dynamics in a humic lake: differential persistence of common freshwater phylotypes. Environ Microbiol 8: 956–70. doi: 10.1111/j.1462-2920.2005.00979.x

63. Liu Z, Huang S, Sun G, Xu Z, Xu M (2012) Phylogenetic diversity, composition and distribution of bacterioplankton community in the Dongjiang River, China. FEMS Microb Ecol 80: 30–44. doi: 10.1111/j.1574-6941.2011.01268.x

64. Crump BC, Armbrust EV, Baross JA (1999) Phylogenetic analysis of particle-attached and free-living bacterial communities in the Columbia river, its estuary, and the adjacent coastal ocean. Appl Environ Microbiol 65: 3192–3204.

65. Venter JC, Remington K, Heidelberg JF, Halpern AL, Rusch D, et al. (2004) Environmental Genome Shotgun Sequencing of the Sargasso Sea. Science 304: 66–74. doi: 10.1126/science.1093857

66. Burkert U, Warnecke F, Babenzien D, Zwirnmann E, Pernthaler J (2003) Members of a readily enriched beta-proteobacterial clade are common in surface waters of a humic lake. Appl Environ Microbiol 69: 6550–6559. doi: 10.1128/aem.69.11.6550-6559.2003

67. Simek K, Hornak K, Jezbera J, Masin M, Nedoma J, et al. (2005) Influence of top-down and bottom-up manipulations on the R-BT065 subcluster of β-Proteobacteria an abundant group in bacterioplankton of a freshwater reservior. Appl Environ Microbiol 71: 2381–2390. doi: 10.1128/aem.71.5.2381-2390.2005

68. Warnecke F, Amann R, Pernthaler J (2004) Actinobacterial 16S rRNA genes from freshwater habitats cluster in four distinct lineages. Environ Microbiol 6: 242–253. doi: 10.1111/j.1462-2920.2004.00561.x

69. Lemke MJ, Lienau EK, Rothe J, Pagioro TA, Rosenfeld J, et al. (2009) Description of freshwater bacterial assemblages from the upper Parana River flood pulse system, Brazil. Microb Ecol 57: 94–103. doi: 10.1007/s00248-008-9398-3

70. Brierley EDR, Wood M (2001) Heterotrophic nitrification in an acid forest soil: isolation and characterisation of a nitrifying bacterium. Soil Biol Biochem 33: 1403–1409. doi: 10.1016/s0038-0717(01)00045-1

71. Elifantz H, Malmstrom RR, Cottrell MT, Kirchman DL (2005) Assimilation of polysaccharides and glucose by major bacterial groups in the Delaware Estuary. Appl Environ Microbiol 71: 7799–7805. doi: 10.1128/aem.71.12.7799-7805.2005

72. Glockner FO, Zaichikov E, Belkova N, Denissova L, Pernthaler J, et al. (2000) Comparative 16S rRNA analysis of lake bacterioplankton reveals globally distributed phylogenetic clusters including an abundant group of Actinobacteria. Appl Environ Microbiol 66: 5053–5065. doi: 10.1128/aem.66.11.5053-5065.2000

Longitudinal Variability of Phosphorus Fractions in Sediments of a Canyon Reservoir Due to Cascade Dam Construction: A Case Study in Lancang River, China

Qi Liu, Shiliang Liu, Haidi Zhao, Li Deng, Cong Wang, Qinghe Zhao, and Shikui Dong

School of Environment, State Key Laboratory of Water Environment Simulation, Beijing Normal University, Beijing, China

ABSTRACT

Dam construction causes the accumulation of phosphorus in the sediments of reservoirs and increases the release rate of internal phosphorus (P) loading. This study investigated the longitudinal variability of phosphorus fractions in sediments and the relationship between the contents of phosphorus fractions and its influencing factors of the Manwan Reservoir, Lancang River, Yunnan Province, China. Five sedimentary phosphorus fractions were quantified separately: loosely bound P (ex-P); reductant soluble P (BD-P); metal oxide-bound P (NaOH-P); calcium-bound P (HCl-P), and residual-P. The results showed that the total phosphorus contents ranged from 623 to 899 µg/g and were correlated positively with iron content in the sediments of the reservoir. The rank order of P fractions in sediments of the mainstream was HCl-P>NaOH-P>residual-P>BD-P>ex-P, while it was residual-P>HCl-P>NaOH-P>BD-P>ex-P in those of the tributaries. The contents of bio-available phosphorus in the tributaries, including ex-P, BD-P and NaOH-P, were significantly lower than those in the mainstream. The contents of ex-P, BD-P, NaOH-P showed a similar increasing trend from the tail to the head of the Manwan Reservoir, which contributed to the relatively higher content of bio-available phosphorus, and represents a high bio-available phosphorus releasing risk within a distance of 10 km from Manwan Dam. Correlation and redundancy analyses showed that distance to Manwan Dam and the silt/clay fraction of sediments were related closely to the spatial variation of bio-available phosphorus.

INTRODUCTION

Dam construction can change rivers' configurations and flow regimes [1], [2], [3], which will have conspicuous direct effects on nutrient loading in the rivers [4], [5]. For example, after the closure of the Three-Gorges Dam in China, the nutrient concentrations and ratios in the water declined dramatically during the high flood season [4]. In the upper stream of the Yellow River, nutrient pollutants varied greatly because of the construction of cascade dams [6]. Dams reduce the transportation of nutrients to marine waters, which indicates that more nutrients settle in the sediments [5].

Among different nutrient pollutants, phosphorus (P) has attracted much attention in sediment research as a key nutrient for phytoplankton growth, controlling the primary productivity of reservoirs [7], [8], [9], [10], [11]. When external loading of P occurs, such as from an increase in the drainage of intensively cultivated areas and from sewage, the rate of accumulation of P in the sediments exceeds its ability to release P into water, so sediments act as P sinks. While the external P is reduced, the sediments still release P into the water, which is called internal phosphorus (P) loading [8], [9], [11], [12], [13], [14], [15], [16]. Previous research has found that, even when the external phosphorus loading was reduced, P concentrations in lakes either did not change or decreased slightly because of internal phosphorus loading [17], [18]. Further, P released by internal P loading contributes to the pool of the P used easily by algae in sediments [19]. Hence, internal P loading has been of great concern in recent years because it can be a potential hazard to aquatic ecosystems [20]. For example, a study in Taihu Lake in China indicated that more than 50% of the inorganic P could be released into the water and used by algae under certain conditions [21]. Although there are several forms of P in sediments, not all of them are released easily from the sediments into the water [22], as this depends on the characteristics of the sediment, environmental factors and the concentration of P in water as well [8], [14], [22], [23], [24]. Studies have shown that grain size distribution of the sediments influences the element composition in the sediments, including metal and nutrient contents [25],[26], [27]. Stone and English (1993) [19] found that fine grained sediment had a different relationship with different P fractions. Further, in river systems influenced by hydropower dams, the spatial grain size distribution of the sediments might be affected by the formative reservoirs and prolonged water renewal time, which eventually has an effect on the heterogeneity of different P fractions in the sediments. Recently, determining P fractions in sediments and the releasing capacity of P fractions from the sediments to the water has been studied and reviewed extensively [12], [28], [29], [30], [31]. However, the study of the mechanism of the longitudinal variability of phosphorus fractions in the sediments of canyon reservoirs affected by cascade dams is insufficient and needs further research.

Studies have shown that, in the Lower Mekong River mainstream, countries through which the Mekong flows are currently threatened by accelerated eutrophication caused by both hydropower development and climate change [32]. Moreover, the Mekong River Commission (MRC) Water Quality Report (2008) noted that almost one-third of the total P in the Lower Mekong Basin exists as soluble orthophosphate (PO_4-P), which indicated that approximately two-thirds of the total P loading was associated with sediments. In the Upper Mekong River, the external loading of P is lower than that of the Lower Mekong River. Therefore, even slight changes in sediment conditions are likely to affect the P concentration of the water column [33]. However, there are few studies regarding the nutrient condition of the sediments in the Upper Mekong section in China (Lancang River), where a chain of fourteen cascade hydroelectric dams have been planned since the early 1980s, with some completed or currently under construction. Further, human activities are becoming more and more intense, with industrialization and fertilization causing an increased P loading in the river. The Manwan Dam was the first dam constructed in the cascade development project along the Lancang River mainstream. After the Manwan Reservoir began operation, siltation led to increasing concern from researchers and the public because it had increased the elevation of the reservoir bottom by as much as 30 m caused by dam construction [34]. Studies have shown that there was great spatial and temporal variation of heavy metals in the sediments of the Manwan Reservoir, and some metals, such as As, Cd, Cr, Cu, Pb and Zn, have reached contamination levels [35], [36]. Further, the study of Zhao, et al. [37], revealed that the potential ecological risk (RI) of multiple heavy metals was related to the grain size of sediments and correlated negatively with the distance from Manwan Dam, a valuable finding for understanding heavy metal contamination in a canyon reservoir. However, the possible spatial variation of P fractions in sediments remains unknown, and this information is useful in understanding the dynamic of trophic conditions of the Manwan Reservoir induced by dam construction.

The aims of our study were to (1) investigate the spatial variation of different P fractions in river sediments of the mainstream and the tributaries of the Manwan Reservoir; (2) estimate the contents of bio-available P in the sediments of Manwan Reservoir, and (3) explore the relationship among P fractions and influencing factors, including metals and sediment grain size.

MATERIALS AND METHODS

Study Site

The Mekong River is the largest international river in Asia, flowing through seven climatic zones and five countries; it is considered to be one of the most important cradles of human civilization in Southeast Asia [38], [39]. The Upper Mekong Section in China (Lancang River), with almost 91% of the drop in elevation of the Mekong River, produces plentiful hydraulic resources [40],[41], [42]. To date, fourteen cascade dams have been planned there since the early 1980s [35], four of which have been constructed (Xiaowan, Manwan, Dachaoshan and Jinghong) in Yunnan Province. The Manwan Dam, completed in 1993, is the first multimillion kilowatt hydropower station in Yunnan Province [39], [42]. The dam is 418 m long and 132 m high with a backwater of 70 km near the Xiaowan Dam. The Manwan Reservoir was a canyon reservoir located in a gorge flanked by high mountains, most of whose peaks are higher than 2,200 m above sea level, and deep valleys with a gradient ratio over 15% [36] (Fig. 1). The area of Manwan Reservoir is 23.6 km^2, and the width of the water surface is, on average, 337.1 m, 2.8 times larger in area and 2 times wider than it was before dam construction. The total reservoir capacity is $1,060 \times 10^6$ m^3, with a normal water level of 994 m; the effective capacity is $257 \times 10^6 m^3$ depending upon seasonal discharge regulation [42].

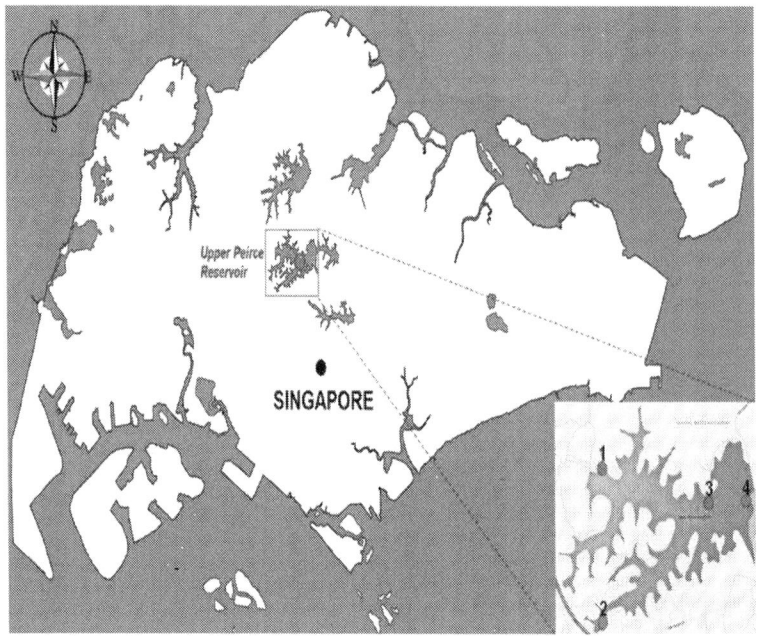

Figure 1: Location of the Manwan Reservoir and the 19 cross-sectional sediment samples, Yunnan Province, China.

Sediment Sample Collection

Manwan Dam was constructed and is managed by China Huaneng Group Corporation, a state-owned enterprise. The corporation gave us permission to conduct this field study, which did not involve any endangered or protected species. It was environmentally neutral and did not threaten the welfare of any species or that of the local population. Therefore, it was not related to ethical issues and no specific permissions were required for such activities.

In June, 2012, nineteen surface (15 cm) sediment samples were collected, using cable operated sediment samplers (Van Veen grabs) to investigate the spatial variation in P fractions in the sediments of the Manwan Reservoir. Fourteen cross-sectional samples (S1–S14) were selected to examine the variations in the mainstream of the reservoir, and the remaining five cross-sectional samples (S15–S19) were located at the major tributaries of the reservoir near the dam (Fig. 1). Each sample in a cross-section was a mixture of three sampling sites: the left, middle and right of each section. Both the right and left sampling sites were selected at the same distance to the shore. All of the samples were placed in sealed plastic bags and maintained at 4°C until analysis. After transportation to the laboratory, the samples were kept frozen, and before the analysis, were freeze-dried and ground until all of the particles passed through a 2 mm nylon sieve after removal of the coarse debris [27], [43], [44].

Analytical Methods

For P fractionation of the sediments from the reservoir, we used the chemical sequential extraction method of Psenner et al. [45], slightly modified by Hupfer et al. [46]. This method fractionates the phosphorus of the sediments into four fractions–loosely bound P (ex-P), reductant soluble P (BD-P), metal oxide-bound P (NaOH-P), calcium-bound P (HCl-P; Table 1) and residual-P–which was the difference between total phosphorus and the four P fractions extracted. All procedures were carried out in triplicate to yield reliable results.

Table 1: Extraction procedure used in this work

step	sequential extraction method	P fraction
1	1 g sediments added to 25 ml 1 M NH_4Cl at pH = 7 shaken for 4 h	ex-P
2	Residual sample added to 0.11 M $Na_2S_2O_4$/$NaHCO_3$* shaken for 1 h at 40uC.	BD-P
3	Residual sample added to 0.1 M NaOH shaken for 16 h	NaOH-P
4	Residual sample added to 0.5 M HCl shaken for 16 h	HCl-P

*Both Na2S2O4 and NaHCO3 were the same concentration. doi:10.1371/journal.pone.0083329.t001

The extracts in every step were centrifuged at 4500 r/min for 20 minutes, and the soluble reactive phosphorus (SRP) in each fraction was determined by the molybdenum blue/ascorbic acid method (APHA, 1985). For the NaOH extracts, the supernatants were filtered through a 0.45-μm poly-amide filter.

The concentration of total phosphorus was determined by ICP-AES after acid digestion of the freeze-dried samples. Residual phosphorus was the difference between total phosphorus and the sum of the four P fractions extracted above [12]. Total concentrations of Ca, Fe, Mn and Al were determined by the SEPAC method (HJ/T 166-2004) using ICP-AES after wet digestion[47]. The results were calculated on the basis of dry weight sediment.

According to the Unified Soil Classification System (USCS), the sediment particles were classified into four grain sizes: coarse/medium sand fraction (246–840 μm); fine sand fraction (147–246 μm); very fine sand fraction (74–147 μm), and silt/clay fraction (<74 μm). The grain size of the sediments was analyzed by an LS 230 laser diffraction particle analyzer (Microtrac S3500).

Statistical Analysis

Redundancy analysis (RDA) is a multivariate direct gradient analysis that enables the identification of variables that best explain the variance pattern of the P fractions [48], [49], [50]. The data were log $(x+1)$ transformed, centered and standardized before a forward selection procedure combined with Monte Carlo permutation tests (499 permutations) were used to identify which factors contributed significantly to the variation ($p<0.05$). The analysis was performed in CANOCO, Version 4 for Windows, and the results are presented in an ordination diagram in which all of the variables are represented by arrows. A smaller angle between arrows represents a high correlation between variables, and the direction of the arrows represents positive or negative correlations. Pearson correlation analyses were used to provide a further quantitative explanation of the correlation between P fractions and metal content. The P fractions contents in the sediments from different sampling sites were subjected to one-way ANOVA to detect significant differences. The correlation and variance analyses were performed in SPSS 18.0.

RESULTS AND DISCUSSION

The Contents of P Fractions and Related Metals in the Manwan Reservoir

The statistical results with respect to different P fractions, related metals and grain size of the sediment particles of the Manwan Reservoir are presented in Table 2. The contents of different P fractions varied greatly. Ex-P is dissolved P, which is absorbed lightly onto the surface of sediment particles or is released from leached P or $CaCO_3$-associated P from organic debris [9], [51]. It can estimate the amounts of phosphorus immediately available. In Manwan Reservoir, ex-P consisted of the minimum part of the P pool (0.1% on average). BD-P is mainly redox sensitive P that binds to Fe-hydroxides and Mn compounds and is considered to be potentially available to algae [9], [28]. Under the anaerobic conditions of the water-sediment interface, BD-P will be released into the water by reductive Fe dissolution. In Manwan Reservoir, BD-P constituted a minor part of total phosphorus (TP) of the sediments (4.9% on average) which was relatively lower than in other reservoirs. Using the same extracting methods in Lake Simcoe, which is the largest lake in South Ontario and is considered mesotrophic (20%–42% TP), Dittrich et al. found that BD-P was the dominant fraction, accounting for 40% and 57%, respectively, of the long- and short-term sediment P release [52]. In Manwan Reservoir, reactive NaOH-P was the third most abundant P fraction in the sediments and accounted for 19.6% of the total phosphorus, on average. Reactive NaOH-P mainly presented as P bound to the surface of aluminium oxides released at high pH levels because of the ligand exchange reactions of hydroxide ions replacing orthophosphate [12], [53], as well as some interior Fe oxides that were not extracted in the BD step [8]. Further, in previous studies, NaOH-P has been used to estimate both short- and long-term available P in the sediments and was verified to be an indicator of algal-available P [54], [55]. HCl-P mainly represents calcium-bound P [56], which appears to be non-motile and is not easily bio-available in the sediments [24], [57], [58]. HCl-P was the most abundant P fraction in the sediments of Manwan Reservoir, constituting 43.6% of TP on average, almost equal to the percentage in Lake Simcoe [52]. Residual-P includes organic phosphorus and refractory P compounds. In Manwan Reservoir, residual-P was the second most abundant P fraction, accounting for 31.9% of TP on average. The rank order of the average concentrations of the four metals in the sediments was Al>Fe>Ca>Mn. The analytical results also showed that there was great spatial variation in the distribution of sediment particles in Manwan Reservoir. According to the average percentage, the silt/clay fraction composed the largest part of all fractions (47.9% on average). The sand fraction, including very fine, fine, and coarse sand, accounted for 45.2% total. Table 2 also shows that Ex-P, Ca, and the coarse sand fraction had relatively high coefficients of variation, which indicated that their contents varied greatly in Manwan Reservoir.

Table 2: Statistical analysis of P fraction concentrations, metals in the sediments and grain size of the sediments of the Manwan Reservoir

	Range (µg/g)	Average percentage (%)	Average concentration (pg/g)	SD	CV
ex-P	0-1.4	0.1	0.5	0.4	0.8
BD-P	8.7-80.4	4.9	35.5	21.7	0.6
NaOH-P	57.3-270.2	19.6	155.3	67.7	0.4
HCl-P	137.5-403.0	43.6	315.6	91.5	0.3
Residual-P	66.2-369.8	31.9	228.8	97.4	0.4
Al	3.89×10^4-7.57×10^4	49.4	6.07×10^4	1.17×10^4	0.2
Ca	2.37×10^3-4.61×10^4	31.6	2.25×10^4	1.14×10^4	0.5
Fe	2.65×10^4-4.31×10^4	18.4	3.71×10^4	4.62×10^4	0.1
Mn	2.71×10^2-1.09×10^3	0.6	6.76×10^2	1.81×10^2	0.3
S/C	-	47.9	-	33.3	0.7
VFS	-	14.5	-	8.8	0.6
FS	-	12.1	-	15.2	1.3
CS	-	18.6	-	29.9	15

S/C: slit/clay fraction (,74 mm).

VFS : very fine sand (74–147 mm).

FS: fine sand (147–246 mm).

CS: coarse sand (246–840 mm).

doi:10.1371/journal.pone.0083329.t002

Spatial Variation of P Fractions in the Mainstream and Tributaries

The relationship between concentrations of P fractions and the distance from the sampling site to Manwan Dam (dis-MW) in the mainstream is presented in Fig. 2. In the mainstream, the changing trends of ex-P, BD-P and NaOH-P were similar from the tail to the head of the Manwan Reservoir. The average content of these three P-fractions increased with decreasing distance to the Manwan Dam. Especially within 9 km from the Manwan Dam, the average content of these three P fractions was relatively higher than it was in other parts of the Manwan Reservoir. Particularly, NaOH-P increased faster than ex-P and BD-P. Unlike ex-P, BD-P and NaOH-P, there was not a consistent pattern in the distribution of HCl-P and TP with respect to the distance to the Manwan Dam; further, there was an upward trend in the concentration of residual-P from Manwan to Xiaowan Dam.

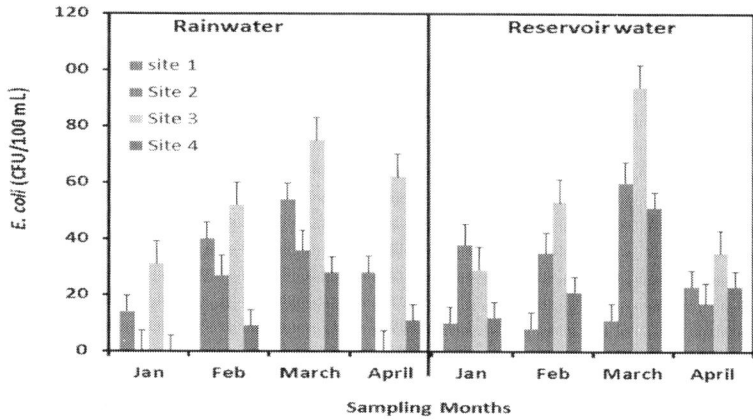

Figure 2: Spatial variation of P fractions in the sediments of the mainstream in Manwan Reservoir.

A previous study has shown that in the Haihe River, China, the amount of P released from sediments was related closely to the ex-P and BD-P fractions, indicating that the P from these two fractions can be released easily [29]. Furthermore, previous studies have estimated the bio-available phosphorus (BAP) in sediments according to the sequential chemical extraction method by the sum of ex-P, BD-P and NaOH-P [21]. By virtue of this theory, Fig. 3 presents the spatial variation of BAP contents in the mainstream of Manwan Reservoir. The values of BAP were at a maximum in the section between S10 and S14, indicating a higher release risk of bio-available P in the sediment. Further, five tributaries were situated within a distance of 10 km from Manwan dam. Samples in the mainstream (S10–S14) and tributaries (S15–S19) within a distance of 10 km to Manwan Dam were selected to compare the spatial variation of P fractions and BAP using a one-way ANOVA (Fig. 4). The results revealed that there were significant differences in the concentrations of ex-P, HCl-P, residual-P and BAP values between the mainstream and the tributaries at the head of Manwan Reservoir. In the mainstream, the average concentrations of ex-P, HCl-P and residual-P were 1.08 µg/g, 380.50 µg/g and 153.47 µg/g, respectively, whereas in the tributaries, the concentrations of ex-P and HCl-P were smaller by 63% and 56%, respectively, than those in the mainstream. Additionally, the average concentration of residual-P was significantly higher in tributaries than in the mainstream. Also in the mainstream, the contents of BAP were significantly higher than those in the tributaries. Because there are more tributaries at the head of Manwan Reservoir, the untreated sewage and drainage of phosphorus from the agricultural land near the tributaries will merge into the mainstream, causing an increase in inorganic P in the sediments.

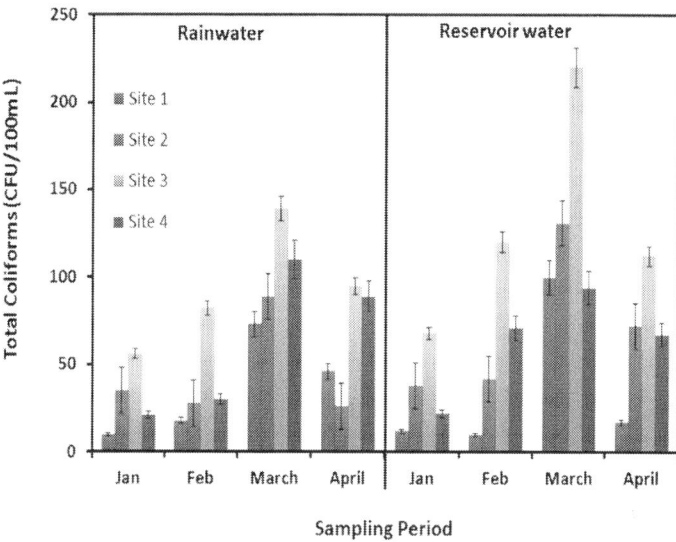

Figure 3: Spatial variation of BAP in the sediments of the mainstream in Man-wan Reservoir.

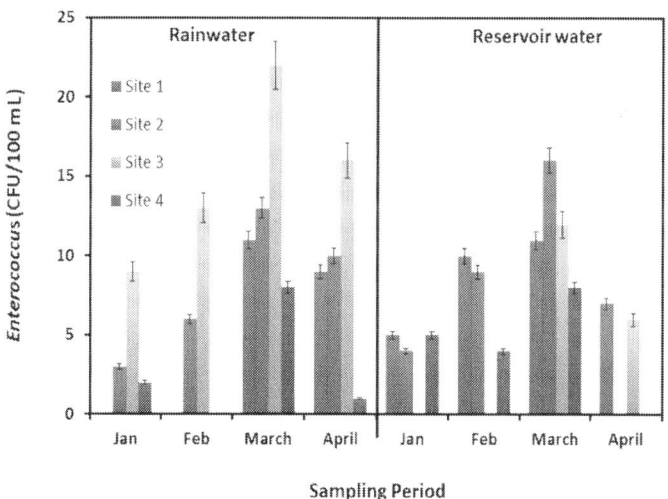

Figure 4: Average concentrations of P fractions in the sediments of mainstream and tributaries.

(Different letters indicate significant differences among the mainstream and the tributaries at the 0.05 level).

The total phosphorus (TP) and the proportion of BAP contents in inorganic P (IP) in Manwan Reservoir compared with other freshwater systems is presented in Table 3. The concentration of TP in Manwan Reservoir ranged from 623–899 µg/g, which was lower than other mesotrophic lakes, indicating that Manwan Reservoir may be low-to-mesotrophic. Therefore, phosphorus released by sediments may influence the water quality greatly. The percentage of BAP in IP indicated that the potential risk of P releasing in the Manwan Reservoir was relatively lower than that in other lakes and rivers. However, constant monitoring and further study are still required to understand the future internal loading in Manwan Reservoir.

Table 3: Comparison of TP and the proportion of BAP in IP in different surface sediments

Sediment source	TP(µg/g)	BAP in IP (%)	Trophic classification	Reference
Lake Koronia, Greece	1156-1305	40-60	hypereutrophic	Christophoros Christophoridis et al, 2006
Lake Volvi, Greece	776-1044	60-80	meso-to-eutrophic	Christophoros Christophoridis et al, 2006
Lake Erken, Sweden	1814	61.4	mesotrophic	Emile Rydin, 2000
Haihe River, China	968-2017	42.2-65.3	mesotrophic	SUN Shujuan et al, 2009
Manwan Reservoir	623-899	37.7		Present study

doi:10.1371/journal.pone.0083329.t003

Relationship between P Fractions and Influencing Factors

Table 4 shows the concentrations of Al, Ca, Fe and Mn in sediments both in the mainstream and the tributaries of the Manwan Reservoir. The rank order of the concentrations of these four metals both in mainstream and tributaries was Al>Fe>Ca>Mn. The Al content in the tributaries was higher than that in the mainstream, whereas the content of Fe, Ca and Mn in the tributaries was 20%, 47% and 48% lower, respectively, than those in the mainstream. Al and Fe are lithogenous materials and are conservative in the migration process, which contributes to the high content of these metals in the sediments [35], [59].

Table 4: P-related metals concentrations in the Manwan Reservoir ($\times 10^4$ **µg/g**)

		Al	Ca	Fe	Mn
Mainstream	Range	3.67-7.15	1.55-4.18	2.98-4.31	0.05-0.11
	Mean	5.97	2.43	3.73	0.07
Tributaries	Range	3.89-7.58	0.22-4.61	2.65-4.21	0.03-0.07
	Mean	6.41	1.65	3.65	0.05

doi:10.1371/journal.pone.0083329.t004

Correlation analysis was used to examine the relationship between the contents of P fractions and related metals (Table 5). The results indicated that there were close relationships among P fractions, metals and the grain size fraction of the sediments. Both ex-P and BAP were correlated positively with the contents of the silt/clay fraction in the sediments; BAP and TP were correlated negatively with the contents of the coarse/medium sand fraction (246–840 µm) and the fine sand fraction (147–246 µm). A significant correlation between the concentration of iron and the silt/clay fraction of sediments was also observed (r = 0.544, $p<0.05$). Previous researchers found that grain size has an effect on the chemical composition of the sediments, including the metal contents and the P sorption-desorption ability [25], [55]. As a result, the silt/clay fraction may contain more elements, such as iron, which are very important in adsorbing different P fractions, especially NaOH-P and BD-P compositions of BAP [12], [60],[61]. Ex-P is another fraction of BAP that includes phosphorus lightly absorbed onto the surface of sediment particles; therefore, it is associated closely with the surface physical characteristics of sediments. As the silt/clay fraction has more surface area, so it can absorb more ex-P. The results were consistent with studies performed in the mainstream of Haihe River and Keelung River in China, where a significant correlation between ex-P and silt/clay fraction was found and iron showed a positive linear relationship with fine-grained sediments (grain size<63 μm) [55],[60]. Distance to Manwan Dam was correlated significantly with the contents of NaOH-P and BAP, and was also correlated with the contents of coarse sand (r = 0.581, p<0.05) and aluminium (r = −0.486, p<0.05) in the sediments (Table 5), indicating that the sediments farther from Manwan Dam contain more coarse sand with fewer aluminium oxides bound to the surface of sediments, which leads to the reduction of NaOH-P and BAP.

Table 5: The correlation analysis among P fractions, metals and grain size of sediments

	ex-P	BD-P	NaOH-P	HCl-P	BAP	TP	S/C	VFS	FS	CS	Al	Ca	Fe	Mn	Dis-MW
ex-P	1														
BD-P		1													
NaOH-P	0.550*		1												
HCl-P				1											
BAP	0.613*		0.942**		1										
TP				0.522*		1									
S/C	0.791**				0.497*		1								
VFS								1							
FS						-0.537*			1						
CS					-0.504*					1					
Al	0.470*										1				

										Ca	Fe	Mn	dis-MW
Ca		—0.524*		-0.494*							1			
Fe					0.501*	0.544*	-0.629**	-0.723**	-0.488*	0.584*		1		
Mn	0.536*				0.529*							0.527*	1	
dis-MW		—0.631"		-0.634**				0.581*	-0.486*					1

**p,0.01; *p,0.05. BAP: bio-available phosphorus FS: fine sand (147–246 mm) CS: coarse sand (246–840 mm). S/C: slit/clay fraction (,74 mm) VFS : very fine sand (74–147 mm) Dis-Manwan: distance to Manwan Dam. doi:10.1371/journal.pone.0083329.t005

Flow variability affects the ecological patterns and processes in river systems, such as the nutrients dynamic [26]. Dams can manipulate the flow regime and consequently, in the upstream of the dam, fine suspended particles are captured and accumulated from the floodplain, while the downstream channel becomes eroded, leading to the coarsening of the sediments [62], [63], [64] Because of the combined effect of Xiaowan and Manwan dams, the flow velocity downstream from Xiaowan Dam is much higher than in the upstream of Manwan Dam, leading to coarser sediments fractions downstream of Xiaowan Dam. As it approaches the Manwan Dam, the river surface widens and the flow velocity slows, resulting in the accumulation of more slit/clay sediment fractions. According to correlation analysis, the contents of total phosphorus (TP) were correlated positively with Fe ($r = 0.501$, $p<0.05$). Previous studies found that there was an apparent relationship between TP and Fe [12], [44]. In the Manwan Reservoir, the HCl-P content was not related to Ca. This finding is consistent with other studies [43], [55], [65], and can be attributed to the different sources of HCl-P, including sedimentary processes, the exchanges between calcium- and iron-bound phosphate [66] and P from fertilizer caused by runoff [27], [43].

Redundancy analysis (RDA) was performed, using four P fractions, bio-available P, and total P as response variables and using metals, distance to Manwan hydropower station and grain size distribution as explanatory variables. The percentage of silt/clay fraction in sediments, distance to Manwan hydropower station and the concentration of Mn in sediments explained the variation significantly ($p = 0.014$, 0.018 and 0.05, respectively; Fig. 5). These three variables accounted for 54% of the total variance, of which the percentage of silt/clay fraction accounted for 20%, distance to Manwan hydropower station explained 19% and the concentration of Mn explained 15%. The percentage of silt/clay fraction correlated positively with the contents of ex-P, BD-P, NaOH-P and BAP in sediments, and the contents of Mn correlated positively with BD-P. RDA analysis also revealed the relationship between P fractions and other factors. Further, RDA can present the relationship between different sampling sites. Concentrations of NaOH-P were correlated positively with ex-P, which was mainly due to its association with the silt/clay fraction of the sediments. The samples from the study sites were distributed in different quarters of the RDA biplot. Samples in the mainstream at the head of the Manwan Reservoir (S10–S14) were related to ex-P, BD-P, BAP and silt/clay fraction of the sediments, which represented a high release risk of P.

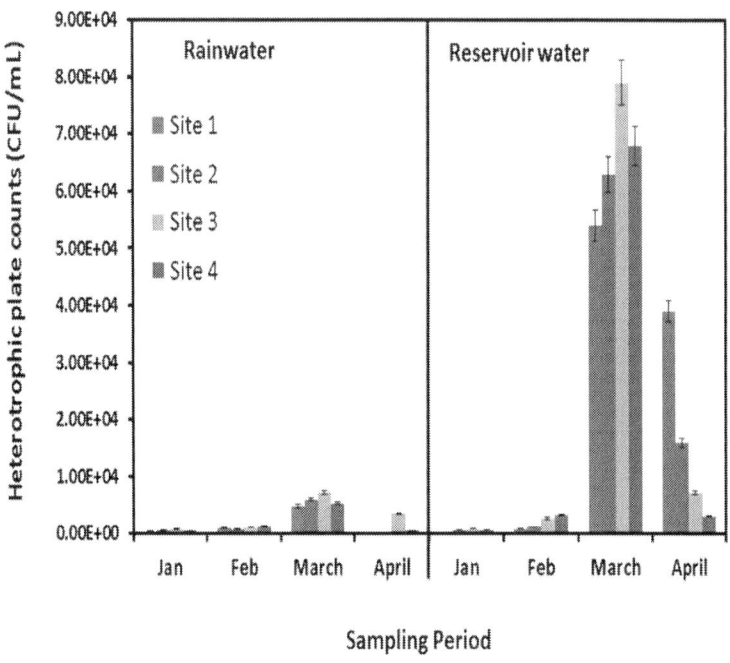

Figure 5: RDA results for P fractions, metals and grain size of sediments.

CONCLUSIONS

Because of the construction of dams, reservoirs have become sensitive regions that have attracted more and more attention in recent years. The release of phosphorus from sediments can influence water quality and water purification capacity, which will endanger the beneficial uses of the river [67]. Thus, spatial analysis of P fractions is important in understanding nutrient pollution and potential ecological risks in the reservoir system. The results of our study are summarized below.

In the mainstream, the rank order of P fractions was HCl-P>NaOH-P>residual-P>BD-P>ex-P. Further, the concentration of ex-P, BD-P and NaOH-P showed an upward trend from Xiaowan to Manwan dam, where the contents of bio-available P (BAP)–the sum of ex-P, BD-P and NaOH-P–were relatively higher from S10 to S14 in the mainstream, especially within approximately 10 km of the Manwan Dam. Moreover, the concentration of BAP in the mainstream was significantly higher than it was in the tributaries; this indicated a greater release risk of phosphorus and more immediately available P in the water. In contrast, there was no trend in the contents of HCl-P and total phosphorus (TP) along the longitudinal direction of Manwan Reservoir. Hence, it may be better to estimate the internal phosphorus loading of the Manwan Reservoir by the contents of BAP rather than TP in the sediments.

In this study, correlation and redundancy analyses (RDA) revealed that silt/clay fraction contents of sediments and distance to Manwan Reservoir influenced the spatial variation of P fractions strongly, especially the bio-available fractions including ex-P, BD-P and NaOH-P. The contents of total phosphorus in the sediments of Manwan Reservoir was 623–899 µg/g, which were lower than some mesotrophic lakes in the world and depended primarily on the contents of iron in the sediments.

The techniques used in this study can be employed widely in freshwater systems and can contribute to the investigation of the tropic conditions, as well as evaluate the potentially mobile P pool stored in canyon reservoir sediments. Especially for an international river like the Mekong nearly half the length of which was in Yunnan Province in China, the construction and operation of cascade dams upstream may cause significant ecological risks downstream [68], risks that are of international concern. Our studies demonstrated that there exist "longitudinal effects" of dams on some phosphorus fractions. Therefore, a management plan was recommended that will intensify the monitoring of the internal P loading of the reservoir, as well as reduce the external P loading that includes the non-point loading from tributaries.

AUTHOR CONTRIBUTIONS

Conceived and designed the experiments: CW QZ SD LD SL QL. Performed the experiments: HZ QL. Analyzed the data: QL. Contributed reagents/materials/analysis tools: QL SL. Wrote the paper: QL SL.

REFERENCES

1. Topping DJ, Rubin DM, Vierra L Jr (2000) Colorado River sediment transport 1. Natural sediment supply limitation and the influence of Glen Canyon Dam. Water Resources Research 36: 515–542. doi: 10.1029/1999wr900285

2. Kileshye Onema JM, Mazvimavi D, Love D, Mul ML (2006) Effects of selected dams on river flows of Insiza River, Zimbabwe. Physics and Chemistry of the Earth, Parts A/B/C 31: 870–875. doi: 10.1016/j.pce.2006.08.022

3. Wei G, Yang Z, Cui B, Li B, Chen H, et al. (2008) Impact of dam construction on water quality and water self-purification capacity of the Lancang River, China. Water Resources Management 23: 1763–1780. doi: 10.1007/s11269-008-9351-8

4. Gong G-C, Chang J, Chiang K-P, Hsiung T-M, Hung C-C, et al.. (2006) Reduction of primary production and changing of nutrient ratio in the East China Sea: Effect of the Three Gorges Dam? Geophysical Research Letters 33.

5. Warner RF (2012) Environmental impacts of hydroelectric power and other anthropogenic developments on the hydromorphology and ecology of the Durance channel and the Etang de Berre, southeast France. Journal of Environmental Management 104: 35–50. doi: 10.1016/j.jenvman.2012.03.011

6. Ouyang W, Hao F, Song K, Zhang X (2010) Cascade Dam-Induced Hydrological Disturbance and Environmental Impact in the Upper Stream of the Yellow River. Water Resources Management 25: 913–927. doi: 10.1007/s11269-010-9733-6

7. Sanclements MD, Fernandez IJ, Norton SA (2009) Soil and sediment phosphorus fractions in a forested watershed at Acadia National Park, ME, USA. Forest Ecology and Management 258: 2318–2325. doi: 10.1016/j.foreco.2009.03.016

8. Christophoridis C, Fytianos K (2006) Conditions affecting the release of phosphorus from surface lake sediments. Journal of Environment Quality 35: 1181–1192. doi: 10.2134/jeq2005.0213

9. Kaiserli A, Voutsa D, Samara C (2002) Phosphorus fractionation in lake sediments – Lakes Volvi and Koronia, N. Greece. Chemosphere 46: 1147–1155. doi: 10.1016/s0045-6535(01)00242-9

10. Vollenweider RA (1982) Eutrophication of waters: Monitoring, assessment and control: Organisation for Economic Co-operation and Development.

11. Farmer JG, Bailey-Wattes AE, Kirika A, Scott C (1994) Phosphorus fractionation and mobility. Aquatic Conservation: Marine And Freshwater Ecosystem 4: 45–56. doi: 10.1002/aqc.3270040105

12. Fytianos K, Kotzakioti A (2005) Sequential fractionation of phosphorus in lake sediments of Northern Greece. Environmental Monitoring and Assessment 100: 191–200. doi: 10.1007/s10661-005-4770-y

13. Ramm K, Scheps V (1997) Phosphorus balance of a polytrophic shallow lake with consideration of phosphorus release. Hydrobiologia 342/343: 43–53. doi: 10.1007/978-94-011-5648-6_5

14. Zhang B, Fang F, Guo J, Chen Y, Li Z, et al. (2012) Phosphorus fractions and phosphate sorption-release characteristics relevant to the soil composition of water-level-fluctuating zone of Three Gorges Reservoir. Ecological Engineering 40: 153–159. doi: 10.1016/j.ecoleng.2011.12.024

15. Jacoby JM, Lynch DD, Welch EB, Perkins MA (1982) Internal phosphorus loading in a shallow eutrophic lake. Water Research 16: 911–919. doi: 10.1016/0043-1354(82)90022-7

16. Cooke GD, McComas MR, Waller DW, Kennedy RH (1977) The occurrence of internal phosphorus loading in two small, eutrophic, glacial lakes in northeastern Ohio. Hydrobiologia 56: 129–135. doi: 10.1007/bf00023351

17. Boström B, Andersen JM, Fleischer S, Jansson M (1988) Exchange of phosphorus across the sediment-water interface. Hydrobiologia 170: 229–244. doi: 10.1007/bf00024907

18. Spears BM, Carvalho L, Perkins R, Kirika A, Paterson DM (2006) Spatial and historical variation in sediment phosphorus fractions and mobility in a large shallow lake. Water Research 40: 383–391. doi: 10.1016/j.watres.2005.11.013

19. Stone M, English M (1993) Geochemical composition, phosphorus speciation and mass transport of fine-grained sediment in two Lake Erie tributaries. Hydrobiologia 253: 17–29. doi: 10.1007/bf00050719

20. Kopá ek J, Borovec J, Hejzlar J, Ulrich K-U, Norton SA, et al. (2005) Aluminum Control of Phosphorus Sorption by Lake Sediments. Environmental Science & Technology 39: 8784–8789. doi: 10.1021/es050916b

21. Zhou A, Wang D, Hongxiao T (2005) Phosphorus fraction and bioavailability in Taihu Lake(China) sediment. Journal of Environmental Science 17: 384–388.

22. Xiang S, Zhou W (2011) Phosphorus forms and distribution in the sediments of Poyang Lake, China. International Journal of Sediment Research 26: 230–238. doi: 10.1016/s1001-6279(11)60089-9

23. Danen-Louwerse H, Lijklema L, Coenraats M (1993) Iron content of sediment and phosphate adsorption properties. Hydrobiologia 253: 311–317. doi: 10.1007/bf00050751

24. Jin X, Wang S, Pang Y, Chang Wu F (2006) Phosphorus fractions and the effect of pH on the phosphorus release of the sediments from different trophic areas in Taihu Lake, China. Environmental Pollution 139: 288–295. doi: 10.1016/j.envpol.2005.05.010

25. Zhang C, Wang L, Li G, Dong S, Yang J, et al. (2002) Grain size effect on multi-element concentrations in sediments from the intertidal flats of Bohai Bay, China. Applied Geochemistry 17: 59–68. doi: 10.1016/s0883-2927(01)00079-8

26. Maazouzi C, Claret C, Dole-Olivier M-J, Marmonier P (2013) Nutrient dynamics in river bed sediments: effects of hydrological disturbances using experimental flow manipulations. Journal of Soils and Sediments 13: 207–219. doi: 10.1007/s11368-012-0622-x

27. Andrieux-Loyer F, Aminot A (2001) Phosphorus forms related to sediment grain size and geochemical characteristics in French

coastal areas. Estuarine, Coastal and Shelf Science 52: 617–629. doi: 10.1006/ecss.2001.0766

28. Wang S, Jin X, Zhao H, Wu F (2006) Phosphorus fractions and its release in the sediments from the shallow lakes in the middle and lower reaches of Yangtze River area in China. Colloids and Surfaces A: Physicochemical and Engineering Aspects 273: 109–116. doi: 10.1016/j.colsurfa.2005.08.015

29. Sun S, Huang S, Sun X, Wen W (2009) Phosphorus fractions and its release in the sediments of Haihe River, China. Journal of Environmental Sciences 21: 291–295. doi: 10.1016/s1001-0742(08)62266-4

30. Hong Y, Geng J, Qiao S, Zhang Y, Ding L, et al. (2010) Phosphorus fractions and matrix-bound phosphine in coastal surface sediments of the Southwest Yellow Sea. Journal of Hazardous Materials 181: 556–564. doi: 10.1016/j.jhazmat.2010.05.049

31. Ruttenberg KC (1992) Development of a sequential extraction method for different forms of phosphorus in marine sediments. Limnology and Oceanography 37: 1460–1482. doi: 10.4319/lo.1992.37.7.1460

32. Lanza GR (2011) Accelerated Eutrophication in the Mekong River Watershed: Hydropower Development, Climate Change, and Waterborne Disease. Eutrophication: causes, consequences and control. 373–386.

33. Vicente I, Andersen FØ, Hansen HCB, Cruz-Pizarro L, Jensen HS (2010) Water level fluctuations may decrease phosphate adsorption capacity of the sediment in oligotrophic high mountain lakes. Hydrobiologia 651: 253–264. doi: 10.1007/s10750-010-0304-x

34. He D-m, Zhao W-j, Chen L-h (2004) The ecological changes in Manwan reservoir area and its causes Journal-Yunnan University Natural Sciences. 26: 220–226 (in Chinese)..

35. Wang C, Liu S, Zhao Q, Deng L, Dong S (2012) Spatial variation and contamination assessment of heavy metals in sediments in the Manwan Reservoir, Lancang River. Ecotoxicology and Environmental Safety 82: 32–39. doi: 10.1016/j.ecoenv.2012.05.006

36.   Zhao Q, Liu S, Deng L, Yang Z, Dong S, et al. (2012) Spatio-temporal variation of heavy metals in fresh water after dam construction: a case study of the Manwan Reservoir, Lancang River. Environmental Monitoring and Assessment 184: 4253–4266. doi: 10.1007/s10661-011-2260-y

37.   Zhao Q, Liu S, Deng L, Dong S, Wang C (2012) Longitudinal distribution of heavy metals in sediments of a canyon reservoir in Southwest China due to dam construction. Environmental Monitoring and Assessment: 1–10.

38.   Liu S, Cui B, Dong S, Yang Z, Yang M, et al. (2008) Evaluating the influence of road networks on landscape and regional ecological risk–A case study in Lancang River Valley of Southwest China. Ecological Engineering 34: 91–99. doi: 10.1016/j.ecoleng.2008.07.006

39.   Zhao Q, Liu S, Deng L, Dong S, Cong, et al (2012) Landscape change and hydrologic alteration associated with dam construction. International Journal of Applied Earth Observation and Geoinformation 16: 17–26. doi: 10.1016/j.jag.2011.11.009

40.   Heng L, Jiufu L, Haixing T (1998) Present and future of water resources development in Lancang River basin in Yunnan Province. Advances in Water Science 9: 70–76.

41.   He D, Wu S, Peng H, Yang Z, Ou X, et al. (2005) A study of ecosystem changes in longitudinal Range-Gorge region and transboundary eco-security in southwest China. Advances in Earth Science 20: 338–344.

42.   Fu KD, He DM, Lu XX (2008) Sedimentation in the Manwan reservoir in the Upper Mekong and its downstream impacts. Quaternary International 186: 91–99. doi: 10.1016/j.quaint.2007.09.041

43.   Wang P, He M, Lin C, Men B, Liu R, et al. (2009) Phosphorus distribution in the estuarine sediments of the Daliao river, China. Estuarine, Coastal and Shelf Science 84: 246–252. doi: 10.1016/j.ecss.2009.06.020

44.   Lin C, Wang Z, He M, Li Y, Liu R, et al. (2009) Phosphorus sorption and fraction characteristics in the upper, middle and low reach sediments of the Daliao river systems, China. Journal of Hazardous Materials 170: 278–285. doi: 10.1016/j.jhazmat.2009.04.102

45. Psenner R, Pucsko R, Sage M (1984) Fractionation of Organic and Inorganic Phosphorus Compounds in Lake Sediments, An Attempt to Characterize Ecologically Important Fractions (Die Fraktionierung Organischer und Anorganischer Phosphorverbindungen von Sedimenten, Versuch einer Definition Okologisch Wichtiger Fraktionen). Archiv fur Hydrobiologie 1.

46. Hupfer M, Gächter R, Giovanoli R (1995) Transformation of phosphorus species in settling seston and during early sediment diagenesis. Aquatic Sciences-Research Across Boundaries 57: 305–324. doi: 10.1007/bf00878395

47. SEPAC (2004) The technical specification for soil environmental monitoring. Beijing: Environmental Press of China.

48. Gonzalez C, Orellana L, Casanovas S, Pignata M (1998) Environmental conditions and chemical response of a transplanted lichen to an urban area. Journal of environmental management 53: 73–81. doi: 10.1006/jema.1998.0194

49. Birks H (1995) Quantitative palaeoenvironmental reconstructions. Statistical modelling of quaternary science data Technical guide 5: 161–254.

50. Van Dobben H, Wolterbeek HT, Wamelink G, Ter Braak C (2001) Relationship between epiphytic lichens, trace elements and gaseous atmospheric pollutants. Environmental Pollution 112: 163–169. doi: 10.1016/s0269-7491(00)00121-4

51. Pettersson K (2001) Phosphorus characteristics of settling and suspended particles in Lake Erken. Science of the total environment 266: 79–86. doi: 10.1016/s0048-9697(00)00737-3

52. Dittrich M, Chesnyuk A, Gudimov A, McCulloch J, Quaizi S, et al. (2012) Phosphorus retention in a mesotrophic lake under transient loading conditions: Insights from a sediment phosphorus binding form study. Water Research 17: 1433–1447. doi: 10.1016/j.watres.2012.12.006

53. Kozerski HP, Kleeberg A (2007) The Sediments and Benthic-Pelagic Exchange in the Shallow Lake Müggelsee (Berlin, Germany). International Review of Hydrobiology 83: 77–112. doi: 10.1002/iroh.19980830109

54. Zhou Q, Gibson CE, Zhu Y (2001) Evaluation of phosphorus bioavailability in sediments of three contrasting lakes in China

and the UK. Chemosphere 42: 221–225. doi: 10.1016/s0045-6535(00)00129-6

55.   Han L, Huang S, Stanley C, Osborne T (2011) Phosphorus Fractionation in Core Sediments from Haihe River Mainstream, China. Soil and Sediment Contamination 20: 30–53. doi: 10.1080/15320383.2011.528469

56.   Zhang Z, Wang Z, Joseph H, Xu X, Wang H, et al. (2012) The release of phosphorus from sediment into water in subtropical wetlands: a warming microcosm experiment. Hydrological Processes 26: 15–26. doi: 10.1002/hyp.8105

57.   Ribeiro DC, Martins G, Nogueira R, Cruz JV, Brito AG (2008) Phosphorus fractionation in volcanic lake sediments (Azores – Portugal). Chemosphere 70: 1256–1263. doi: 10.1016/j.chemosphere.2007.07.064

58.   Psenner R, Pucsko R (1988) Phosphorus fractionation: advantages and limits of the method for the study of sediment P origins and interactions. Arch Hydrobiol Beih Ergebn Limnol 30: 43–59.

59.   Rubio B, Nombela M, Vilas F (2000) Geochemistry of major and trace elements in sediments of the Ria de Vigo (NW Spain): an assessment of metal pollution. Marine Pollution Bulletin 40: 968–980. doi: 10.1016/s0025-326x(00)00039-4

60.   Huang K-M, Lin S (2003) Consequences and implication of heavy metal spatial variations in sediments of the Keelung River drainage basin, Taiwan. Chemosphere 53: 1113–1121. doi: 10.1016/s0045-6535(03)00592-7

61.   Pan K, Wang W-X (2012) Trace metal contamination in estuarine and coastal environments in China. Science of the total environment 421: 3–16. doi: 10.1016/j.scitotenv.2011.03.013

62.   Luo X, Yang S, Zhang J (2012) The impact of the Three Gorges Dam on the downstream distribution and texture of sediments along the middle and lower Yangtze River (Changjiang) and its estuary, and subsequent sediment dispersal in the East China Sea. Geomorphology 179: 126–140. doi: 10.1016/j.geomorph.2012.05.034

63.   Heath S, Plater A (2010) Records of pan (floodplain wetland) sedimentation as an approach for post-hoc investigation of the hydrological impacts of dam impoundment: The Pongolo river,

KwaZulu-Natal. Water Research 44: 4226–4240. doi: 10.1016/j. watres.2010.05.026

64. Williams GP, Wolman MG (1984) Downstream effects of dams on alluvial rivers. Washington, DC, USA: US Government Printing Office.

65. Morse JW, Cook N (1978) The distribution and form of phosphorus in North Atlantic Ocean deep-sea and continental slope sediments. Limnology and Oceanography 23: 825–830. doi: 10.4319/lo.1978.23.4.0825

66. Golterman H (1988) The calcium-and iron bound phosphate phase diagram. Hydrobiologia 159: 149–151. doi: 10.1007/bf00014722

67. Kim LH, Choi E, Stenstrom MK (2003) Sediment characteristics, phosphorus types and phosphorus release rates between river and lake sediments. Chemosphere 50: 53–61. doi: 10.1016/s0045-6535(02)00310-7

68. Jacobs JW (2002) The Mekong River Commission: transboundary water resources planning and regional security. The Geographical Journal 168: 354–364. doi: 10.1111/j.0016-7398.2002.00061.x

10

1, 4-Hydroquinone is a Hydrogen Reservoir for Fuel Cells and Recyclable via Photocatalytic Water Splitting

Thorsten Wilke, Michael Schneider, and Karl Kleinermanns

Institute of Physical Chemistry, Heinrich-Heine-University, Duesseldorf, Germany

ABSTRACT

Photocatalytic splitting of water was carried out in a two-phase system. Nanocrystalline titanium dioxide was used as photocatalyst and potassium hexacyanoferrate (III)/ (II) as electron transporter. Generated hydrogen was chemically stored by use of a 1, 4-benzoquinone/1, 4-hydroquinone system, which was used as a recyclable fuel in a commercialised direct methanol fuel cell (DMFC). The electrical output of the cell was about half compared to methanol. The conversion process for water splitting and recombination in a fuel cell was monitored by

UV-Vis spectroscopy and compared to a simulated spectrum. Products of side reactions, which lead to a decrease of the overall efficiency, were identified based on UV-Vis investigations. A proof of principle for the use of quinoide systems as a recyclable hydrogen storage system in a photocatalytic water splitting and fuel cell cyclic process was given.

INTRODUCTION

By the year 2050 the total energy consumption is expected to double, as the world's population is steadily increasing. The fossil fuels are not able to meet this energy demand in the long term. Therefore, renewable energy resources will come more sharply into focus. The most promising alternative is solar light, because the amount of energy that arrives on earth every hour from the sun is greater than the amount that is required by the entire humanity in one year [1]. Yet, there is no practical way to transform and store this huge amount of energy efficiently, because the widely used silicon solar cells are of limited use due to their high production costs. Therefore, it is necessary to look for less expensive and in sufficient quantities available alternatives to high-purity silicon. Storage of solar energy is possible for example by batteries and capacitors, but, compared to chemical bonds, these storage systems feature a low energy density. In this regard hydrogen is a good energy reservoir. It can be used as fuel for vehicles or can be converted into electrical energy by the use of fuel cells. The generation of hydrogen by electrolysis requires electrical energy, which could be obtained by use of solar cells, but the effectiveness is just approximately 8% for large-scale facilities [2]. Thus, direct photolytic water splitting by the use of suitable and inexpensive nanocrystalline semiconductors would be a promising alternative. Here the water is split with high efficiency by solar light [1, 3, and 4].

The semiconductor titanium dioxide has a band gap of about 3.1 eV and its conduction band potential is high enough for water splitting [5]. By absorption of photons electrons can be promoted to the excited state and electron-hole pairs (e^- + h^+) are generated, which diffuse separately on the surface of the TiO_2 particles:

$$TiO_2 + h\nu \rightarrow e^- + h^+$$

$$(1)$$

The formed holes in the valence band are able to oxidize molecules, for example water:

$$H_2O + 2h^+ \rightarrow \frac{1}{2}O_2 + 2H^+$$

(2)

The electrons in the conduction band can reduce H^+ to hydrogen as the reduction potential of TiO_2 is sufficiently negative (-0.65 V [6]):

$$2H^+ + 2e^- \rightarrow H_2$$

(3)

For water splitting by TiO_2 we finally obtain the following overall reaction [7]:

$$H_2O + 2h\nu \rightarrow \frac{1}{2}O_2 + H_2$$

(4)

Depending on the particle size, TiO_2 nanoparticles (NPs) absorb at wavelengths smaller than 350 - 380 nm [8]. Larger particles show smaller band gaps, thus the absorption is red-shifted.

For an efficient and safe chemical storage of the generated hydrogen quinoid systems are suitable. They mimic natural processes, e.g., photosynthesis, which also uses quinoid systems like plastoquinone for hydrogen transfer [7].

Substituted quinoides like 2, 3-dichloro-5, 6-dicyano-1, and 4-benzoquinone (DDQ) are known from literature as good hydrogen acceptors [9] and were already investigated by our group in the past [4]. 1, 4-benzoquinone is less efficient compared to substituted quinoides, but in contrary to DDQ it can be converted in a direct methanol fuel cell (DMFC), because of its relatively good resistance to water. Benzoquinone and hydroquinone are well distinguishable by UV-VIS spectroscopy allowing rather easy quantitative analysis.

Our group presented in 2012 a photocatalytic reaction system for splitting water by use of semiconductor nanoparticles [4], which is a further development of an experimental setup introduced by Matsumura et al. in 1999 [9]. The experimental setup consists of a two-phase system. The photocatalytic water splitting takes place in the aqueous phase containing the photocatalyst and the electron transporter. It is covered with a solution of quinone in n-butyronitrile, which forms the organic layer and serves as hydrogen storage system.

Hydroquinone is formed by reduction of benzoquinone and acceptance of two protons.

$$BQ + 2e^- + 2H^+ \rightarrow HQ \tag{5}$$

Benzoquinone and hydroquinone form quinhydrone complexes [BQ·HQ].

$$BQ + HQ \rightarrow [BQ \cdot HQ] \tag{6}$$

Quinhydrone is a dark green 1:1 charge-transfer, which is moderately soluble in n-butyronitrile. Dissolved quinhydrone undergoes a consecutive reduction to hydroquinone.

$$[BQ \cdot HQ] + 2e^- + 2H^+ \rightarrow 2HQ \tag{7}$$

In this paper we show that quinones can be used as fuel for fuel cells. Hydroquinone is converted to benzoquinone in an air-breathing fuel cell normally used with methanol. Benzoquinone is converted back to hydroquinone by photocatalytic water splitting. Commercialised direct methanol fuel cells (DMFC) consist of a polymer electrolyte ion exchange membrane embedded between the anode and the cathode. Both electrodes are composed of three layers: a catalytic, diffusion and a backing layer, mostly based on Pt or Pt Ru as catalyst. For a successful transport of oxygen to the surface of the catalyst, a mixture of carbon and polytetrafluoroethylene is used as diffusion layer [10]. In air-breathing DMFCs atmospheric oxygen is used without active blowing components by diffusion through open holes of the cathode [11]. The following platinum catalysed reactions take place [12-14]

Anode:

$$CH_3OH + H_2O \rightarrow CO_2 + 6H^+ + 6e^- \qquad U^0 = 0.043 \text{ V} \tag{8};$$

Cathode:

$$6H^+ + 6e^- + 1.5O_2 \rightarrow 3H_2O \qquad\qquad U^0 = 1.229 \text{ V} \tag{9};$$

Overall reaction:

$$CH_3OH + 1.5O_2 \rightarrow CO_2 + 2H_2O \qquad U^0 = 1.186 \text{ V} \qquad (10).$$

Under real conditions the open circuit voltage is always lower than the theoretical value, because of over potential effects at both electrodes.

The electrooxidation of hydroquinone can be described in similar way [14]

Anode:

$$2HQ \rightarrow 2BQ + 4H^+ + 4e^- \qquad U^0 = 0.699 \text{ V} \qquad (11);$$

Cathode:

$$4H^+ + 4e^- + O_2 \rightarrow 2H_2O \qquad U^0 = 1.229 \text{ V} \qquad (12);$$

Overall reaction:

$$HQ + 0.5O_2 \rightarrow BQ + H_2O \qquad U^0 = 0.530 \text{ V} \qquad (13).$$

Due to over potential effects the observed open circuit voltage is lower than the theoretical value.

EXPERIMENTS

Titanium nanopowder (Aerosil TiO_2 P-25, Evonik), 1, 4-benzoquinone (99.5%, Aldrich), potassium hexacyanoferrate (III) (99%, Aldrich), n-butyronitrile

(purum, ≥99.0%, Fluka) and 1,4-hydroquinone (≥99%, Aldrich) were used without purification or other treatment.

Photocatalytic Water Splitting

The water splitting experiments were carried out in a quartz cuvette of $45 \times 12.5 \times 12.5 \text{ mm}^3$ size. 30 mg TiO_2 nanopowder was dispersed in 16 mL of an aqueous potassium hexacyanoferrate (III) solution (8.0 mM). 2.29 ml of the dispersion was placed in the cuvette and was carefully covered with 0.71 mL of a 1, 4-benzoquinone solution in n-butyronitrile (1.9 mM) to form the organic phase. In order to avoid an evaporation of the organic solvent, the cuvette was sealed with a PTFE plug. To avert heating of the reaction system during irradiation and to reduce diffusion of BQ into the aqueous phase, the vessel was cooled to 15˚C by a water flushed aluminium block, connected to the cuvette by a heat-conductive paste.

The irradiation was carried out for 90 minutes by an 80 W mercury-vapor lamp (Oriel, Germany), which was mildly focused to give 100 mW/cm^2.

To prevent irradiation of the organic phase and photodegradation of BQ and HQ an aluminium mask of adequate size was used.

To exclude contributions to the spectra, which were not due to the process of water splitting, a reference solution of the organic phase was kept in the dark without contact to the aqueous phase during irradiation.

After 90 min of irradiation a sample of the organic phase was taken, diluted 1:25 with n-butyronitrile and analysed by UV-Vis spectroscopy.

Conversion of 1, 4-Hydroquinone to 1, 4-Benzoquinone in a Direct Methanol Fuel Cell (DMFC)

The reaction was carried out in a F111 direct methanol fuel cell purchased from H-TEC Education GmbH with an electrode area of 4 cm^2 and a maximal power output of 20 mW. The cell was operated as an air-breathing fuel cell. Oxygen was obtained from the atmosphere by diffusion and convection.

To verify the given properties a 3% by weight solution of methanol (99.9%, Aldrich) in water was used.

A 3% by weight solution of 1, 4-hydroquinone in water was poured into the cell until the electrode was soaked. The electrical output of the cell was recorded by a VC-840 digital multimeter (Voltcraft, Germany) for 180 minutes reaction time.

Afterwards a sample of the reaction mixture was diluted with water and analysed by UV-Vis spectroscopy.

Furthermore, UV-Vis spectra of aqueous solutions of 1, 4-benzoquinone and 1, 4-hydroquinone in a 1:3 concentration value (BQ:HQ) were taken. A simulated UV-Vis spectrum of a 1:3 mixture of benzoquinone and hydroquinone in water was obtained by mathematical addition of the recorded spectra.

Experimental Equipment

To verify the given properties of the TiO_2 nanopowder transmission electron microscopy (TEM) images were taken.

The particle size distribution was investigated by transmission electron microscopy (TEM) measurements, which were performed with a HITACHI TEM 7500 microscope equipped with a Mega View II camera (Soft Imaging System) at the Max-Planck institute for coal research (MPI Mülheim a.d. Ruhr).

The TEM image of the TiO_2 nanoparticles is presented in Figure 1. An average particle size of 21 nm and a homogeneous size distribution have been found.

Optical absorption spectra of the organic phase were recorded by using a Cary 300 UV-VIS spectrophotometer operated at a resolution of 1 nm.

RESULTS AND DISCUSSION

Our experimental setup is shown in Figure 1. A solution of 1, 4-benzoquinone in n-butyronitrile forms the organic layer and is located above the aqueous phase, in which the water splitting takes place. A $[Fe^{+II}(CN)_6]^{4-}$-$[Fe^{+III}(CN)_6]^{3-}$ redox system transports the electrons between the two phases. The holes in the valence band of titanium dioxide oxidize water, while electrons excited to the conduction band reduce the $[Fe^{+III}(CN)_6]^{3-}$- to $[Fe^{+II}(CN)_6]^{4-}$-ions. The reduced ferrocyanide ions transport electrons to the interface, where benzoquinone is subsequently reduced to semihydroquinone and hydroquinone by accepting electrons from $[Fe^{+II}(CN)_6]^{4-}$ and protons from water. Because of the spatial separation of reduction and oxidation processes in two different phases, electron-hole recombination can be minimized and the oxidation of hydroquinone by TiO_2-holes can be prevented. The redox scheme in Figure 3 confirms the energetic feasibility of this approach.

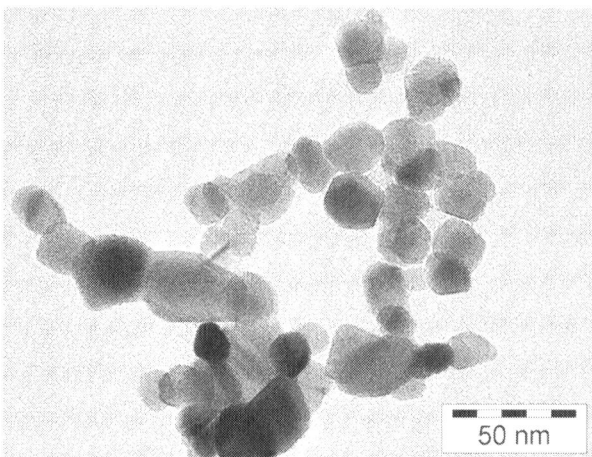

Figure 1: TEM image of TiO$_2$ nanoparticles with an average size of 21 nm obtained from Evonik (Aerosil TiO$_2$ P-25). It consists of a mixture of 80% anastase and 20% rutile with a specific surface area of approximately 50 m^2/g.

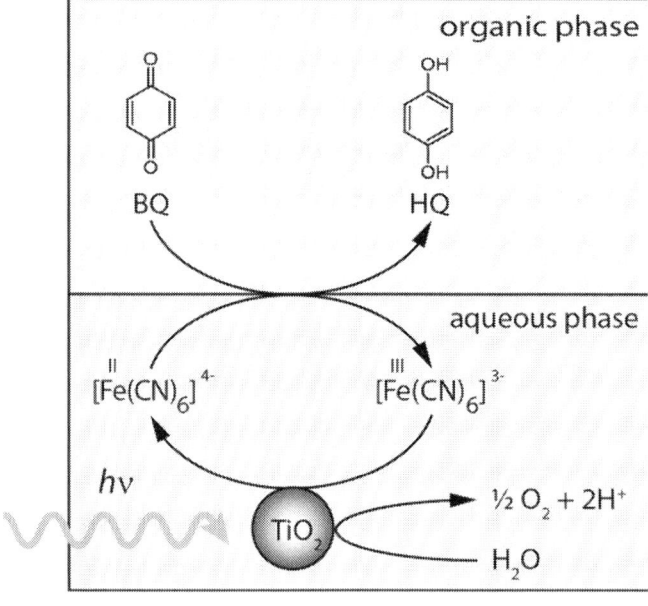

Figure 2: Experimental setup. The reaction is carried out in a two-phase system. Nanocrystalline TiO$_2$ is used as photocatalyst and is suspended in a so-

lution of K_3 [Fe (CN) $_6$] in water, which forms the aqueous phase. It is covered with a layer of the organic phase, a solution of 1, 4-benzoquinone (BQ) in n-butyronitrile [9]. The reaction cell is tempered by a water cooling block to approximately 15°C to reduce diffusion of BQ into the aqueous phase. The aqueous phase is irradiated by a Hg lamp, which is mildly focused to give 100 mW/cm². Aluminium foil prevents irradiation of the organic phase.

Figure 4 shows the UV-Vis spectra of 1, 4-benzoquinone and their suitability for quantitative analysis of the reduction process.

The result of the water splitting experiment is presented in Figure 5. There the UV-Vis spectrum of the diluted organic phase before and after 90 minutes irradiation with the unfiltered Hg lamp spectrum is shown. The strong absorption band at 244 nm can be assigned to benzoquinone. The absorption band at 296 nm, which is much weaker because of different absorption coefficients (1, 4-benzoquinone: $\varepsilon_{244\ nm}$ = 19204 L·mol^{-1}·cm^{-1};

1, 4-hydroquinone: $\varepsilon_{244\ nm}$ = 3453 L·mol^{-1}·cm^{-1}), can be assigned to hydroquinone. A magnification of the hydroquinone absorption band is presented in the inset.

After irradiation for 90 minutes the absorption band at 244 nm decreased by 32% due to the conversion of benzoquinone to hydroquinone and some side reactions, which lead to a loss of quinone. The oxidative polymerisation of benzoquinone in water [15, 16] at the interphase between organic and aqueous phase is the mostimportant side reaction: a small amount of quinone diffuses into the aqueous phase and undergoes condensation reactions leading to high-molecular species of humic acid type [16].

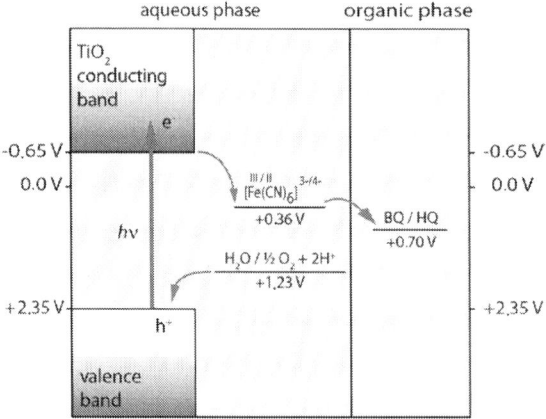

Figure 3: Energy scheme of the two-phase water splitting experiment. Redox potentials are given in Volts [V] against Standard Hydrogen Potential [SHE]. The working principle is based on excitation of TiO_2. The excited electron (e^-) is stored by reducing $[Fe(CN)_6]^{3-}$ to $[Fe(CN)_6]^{4-}$ and transported to the interphase by diffusion. The electron transporter is regenerated to $[Fe(CN)_6]^{3-}$ by reducing BQ. Holes (h^+) in the valence band are filled by oxidation of water leading to O_2 and H^+. HQ is formed at the interface by reaction of H^+ with the reduced BQ.

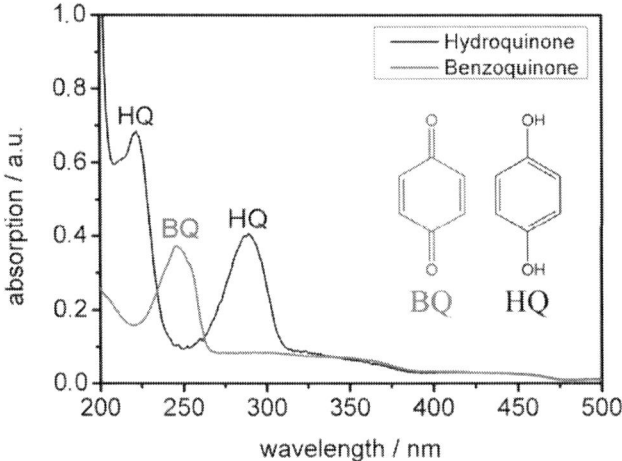

Figure 4: UV-Vis spectra of 1, 4-benzoquinone (red) and 1, 4-hydroquinone (black) measured in water in a 1:5 (BQ:HQ) concentration ratio. Benzoqui-

none shows strong absorption with a maximum at 244 nm, while hydroquinone absorption is considerably weaker with maxima at 222 nm and 296
nm. The spectra were obtained using a suprasil quartz cell with 1 cm path
length and a double-beam UVVis spectrophotometer operated at a resolution
of 1 nm.

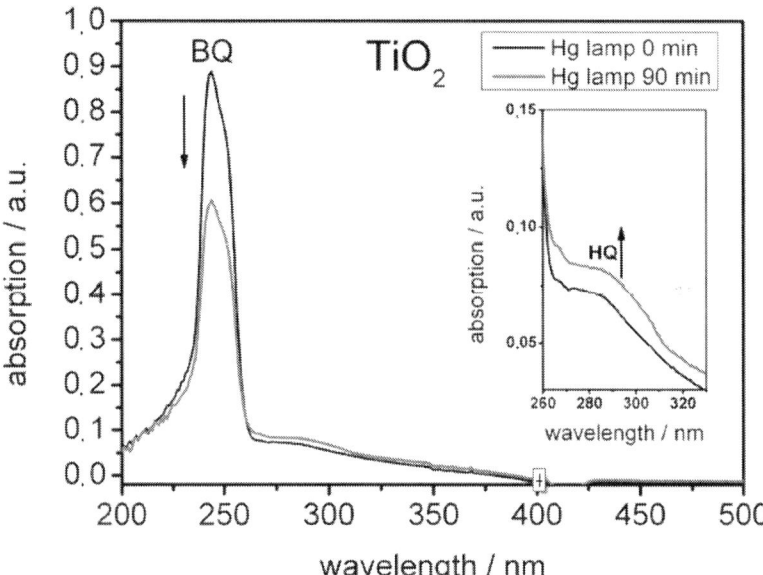

Figure 5: Water splitting by TiO_2 as photocatalyst. Shown are the UV-Vis
spectra of the organic phase before (black line) and after irradiation (red line)
of the aqueous phase with an unfiltered Hg lamp spectrum for 90 minutes.
The recorded strong absorption at 244 nm can be assigned to benzoquinone.
It is decreasing while irradiation due to the conversion of benzoquinone to
hydroquinone. Hydroquinone shows absorption at 296 nm. Because of the
different absorption coefficients the absorption is much weaker compared to
benzoquinone. A magnification of the spectra is shown in the inset.

After 90 minutes of irradiation 37% of the consumed benzoquinone
were converted to hydroquinone. Overall we achieved a total
conversion rate of about 12%. Compared to other quinoide systems
like 2, 3-dichloro5, 6-dicyano-1, 4-benzoquinone (DDQ), which
has been investigated by our group in the past [4], the unsubstituted
p-benzoquinone shows a significantly lower conversion rate. Contrary
to water insoluble 2,3-dichloro- 5,6-dicyano-1,4-hydroquinone

(DDHQ) however, waterbased 1,4-hydroquinone can be converted in a common direct methanol fuel cell (DMFC) into 1,4-benzoquinone. In this way, it is possible to achieve a cyclic process of photocatalytic water splitting for storage of solar energy and a fuel cell to convert it into electric energy. An image of the fuel cell after hydroquinone conversion is presented in Figure 6. To verify the given data, a solution of 3% by weight of methanol in water was used. After 10 minutes reaction time an open circuit voltage of 500 mV and an electrical output of 20 mW was observed. For the conversion of a 3% by weight solution of 1, 4-hydroquinone in water an open circuit voltage of 300 mV and an electrical output of 9 mW was observed. The reaction was carried out for 180 minutes under load. During operation hydroquinone is converted into benzoquinone.

Aside quinhydrone is formed (equation (6)) and some oxidative polymerisation of benzoquinone occurs. Quinhydrone is only poorly soluble in water. Therefore, precipitation of quinhydrone leads to a loss of quinone. An increasing greenish-brown colouring of the reaction solution and a fine precipitation was observed, which is shown in Figure 5.

The UV-Vis spectrum of the reaction mixture after conversion for 180 minutes in the fuel cell is presented in Figure 7 (black line). Two recorded absorption bands at 222 nm and 296 nm can be assigned to unreacted hydroquinone. Only weak absorption of benzoquinone at 244 nm was not observed because of the overlap with the absorption band of hydroquinone and underlying broad absorption of quinhydrone and polymerised benzoquinone. A difference spectrum of experiment and simulation is presented in the inset of Figure 7.

The absorption of quinhydrone in water exhibits an absorption band at 420 nm [17], which can be found in the spectrum of the reaction solution as well. Condensation of benzoquinone leads to a wide variety of highmolecular products of humic acid type, which show a wide absorption in the range of 300nm to 500 nm [16].

CONCLUSIONS

We presented an experimental setup for photocatalytic water splitting by nanocrystalline titanium dioxide and a chemical storage of the generated hydrogen by hydroquinone. The benzoquinone/

hydroquinone system seems to be a good alternative to the formation of gaseous H_2 as it is easy to handle and can be used as a fuel for fuel cells. This work is a proof of principle for the use of quinoide systems as a recyclable storage medium in a photocatalytic water splitting and fuel cell cyclic process. Further work has to be directed to the development of quiniod derivates, which are water soluble and chemically resistant.

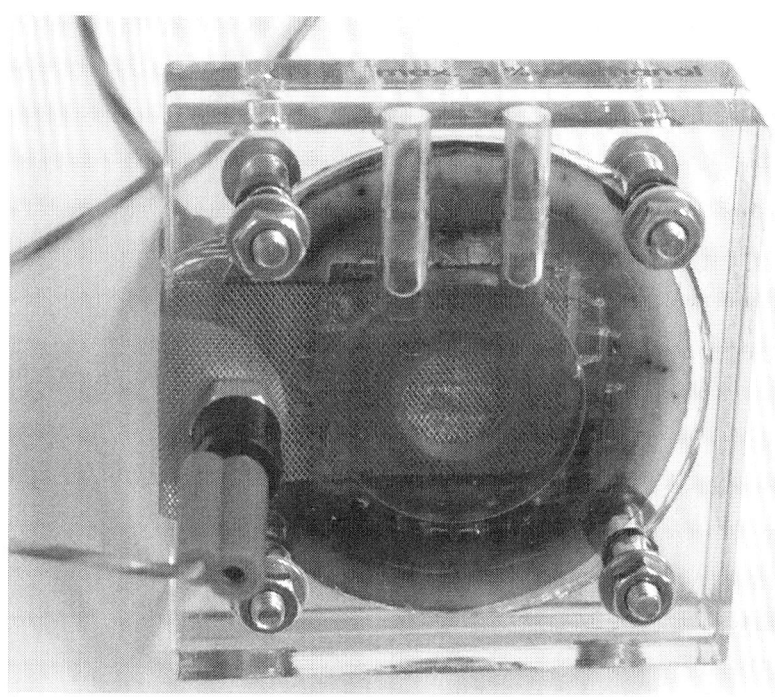

Figure 6: Image of the fuel cell after 180 minutes reaction time. An open circuit voltage of 300 mV, a short circuit current of 30 mA and corresponding electrical power of 9 mW was observed, which is nearly half the methanol fuel cell power An increasing greenish-brown colouring of the reaction solution and a fine precipitate is observed with reaction time due to the formation of quinhydrone and oxidative polymerized benzoquinone [16].

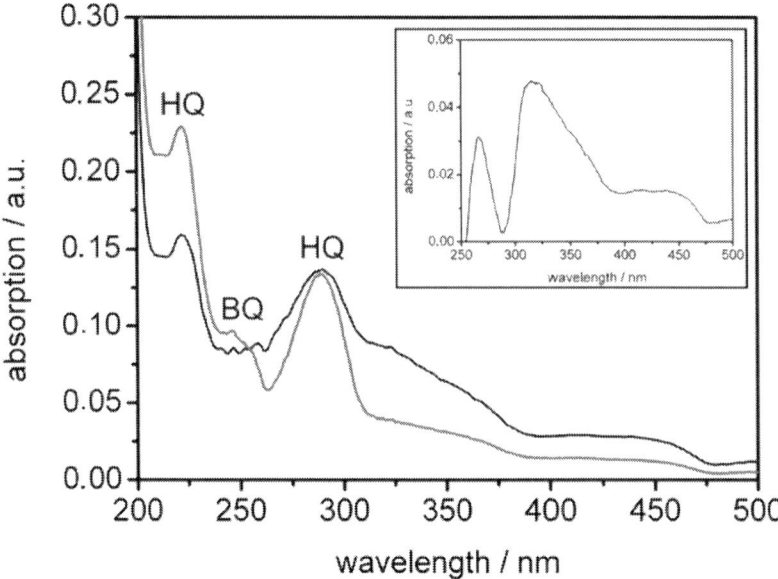

Figure 7: Conversion of hydroquinone in a common direct methanol fuel cell (DMFC). The black line shows the UVVis spectrum of an aqueous solution of hydroquinone (3% in mass) after conversion in a DMFC for 3 hours. The achieved cell power was about half compared to methanol (3% in mass) as fuel. The red line shows a simulated UVVis spectrum of a 1:3 mixture of benzoquinone and hydroquinone in water, achieved by addition of both spectra in a concentration ratio of 1:3 (BQ:HQ). The recorded absorption bands can be assigned to HQ (222 nm and 296 nm) and BQ (244 nm). The inset shows the difference spectrum of the experimental and the simulated UV-Vis spectrum.

ACKNOWLEDGEMENTS

The authors gratefully acknowledge Dr. Christian W. Lehmann and his group (Max-Planck Institut für Kohlenforschung, Mülheim/Ruhr) for the implementation of the TEM measurements.

REFERENCES

1. R. Pike and P. Earis, "Powering the World with Sunlight," Energy & Environmental Science, Vol. 3, No. 2, 2010, p. 173. doi:10.1039/b924940k

2. N. Kelly, T. Gibson and D. Ouwerkerk, "Generation of High-Pressure Hydrogen for Fuel Cell Electric Vehicles Using Photovoltaic-Powered Water Electrolysis," International Journal of Hydrogen Energy, Vol. 36, No. 24, 2011, pp. 15803-15825. doi:10.1016/j.ijhydene.2011.08.058

3. E. Durgun, S. Ciraci, W. Zhou and T. Yildirim, "Transition-Metal-Ethylene Complexes as High-Capacity Hydrogen-Storage Media," Physical Review Letters, Vol. 97, No. 22, 2006, pp. 1-4. doi:10.1103/PhysRevLett.97.226102

4. T. Wilke, D. Schriker, J. Rolf and K. Kleinermanns, "Solar Water Splitting by Semiconductor Nanocomposites and Hydrogen Storage with Quinoid Systems," Open Journal of Physical Chemistry, Vol. 2, No. 4, 2012, pp. 195-203.doi:10.4236/ojpc.2012.24027

5. M. Grätzel, "Photoelectrochemical Cells," Nature, Vol. 414, No. 6861, 2001, pp. 338-344.doi:10.1038/35104607

6. A. Hagfeldt and M. Grätzel, "Light-Induced Redox Reactions in Nanocrystalline Systems," Chemical Reviews, Vol. 95, No. 1, 1995, pp. 49-68. doi:10.1021/cr00033a003

7. M. Kaneko and I. Okura, "Photocatalysis—Science and Technology," Springer, Heidelberg, 2002.

8. D. Ogermann, T. Wilke and K. Kleinermanns, "CdS_xSe_y/ TiO_2 Solar Cell Prepared with Sintered Mixture Deposition," Open Journal of Physical Chemistry, Vol. 2, No. 1, 2012, pp. 47-57. doi:10.4236/ojpc.2012.21007

9. T. Ohno, K. Fujihara, K. Sarukawa, F. Tanigawa and M. Matsumura, "Splitting of Water by Combining Two Photocatalytic Reactions through a Quinone Compound Dissolved in an Oil Phase," Zeitschrift für Physikalische Chemie, Vol. 213, No. 2,1999, pp. 165-174.doi:10.1524/zpch.1999.213.Part_2.165

10. A. S. Aricò, V. Baglio and V. Antonucci, "Electrocatalysis of Direct Methanol Fuel Cells," Wiley-VCH, Weinheim, 2009.

11. I. Chang, M. Lee, J. Du and S. W. Cha, "Characteristic Behaviors on Air-Breathing Direct Methanol Fuel Cells," International Journal of Precision Engineering and Manufacturing, Vol. 13, No. 7, 2012, pp. 1141-1144. doi:10.1007/s12541-012-0151-y

12. H. Dohle, J. Mergel and D. Stolten, "Heat and Power Management of a Direct-Methanol-Fuel-Cell (DMFC) System," Journal of Power Sources, Vol. 111, No. 2, 2002, pp. 268-282.doi:10.1016/S0378-7753(02)00339-7

13. A. S. Aricò, V. Antonucci and N. Giordano, "Methanol Oxidation on Carbon-Supported Platinum-Tin Electrodes in Sulfuric Acid," Journal of Power Sources, Vol. 50, No. 3, 1994, pp. 295-309. doi:10.1016/0378-7753(94)01908-8

14. D. R. Lide and W. M. Haynes, "CRC Hanbook of Chemistry and Physics," Taylor & Francis, London, 2009.

15. L. L. Houk, S. K. Johnson, J. Feng, R. S. Houk and D. C. Johnson, "Electrochemical Incineration of Benzoquinone in Aqueous Media Using a Quaternary Metal Oxide Electrode in the Absence of a Soluble Supporting Electrolyte," Journal of Applied Electrochemistry, Vol. 28, No. 11, 1998, pp. 1167-1177. doi:10.1023/A:1003439727317

16. F. Cataldo, "On the Structure of Macromolecules Obtained by Oxidative Polymerization of Polyhydroxyphenols and Quinones," Polymer International, Vol. 46, No. 4, 1998, pp. 263-268. doi:10.1002/(SICI)1097-0126(199808)46:4<263::AID-PI983>3.0.CO;2-0

17. K. K. Kalninsh and E. F. Panarin, "Catalytic Hydrogen Transfer in Donor-Acceptor Complexes," Doklady Chemistry, Vol. 437, No. 2, 2011, pp. 82-86.doi:10.1134/S0012500811040021

Citations

CHAPTER 1

Mohammad Ali Ahmadi, "Developing a Robust Surrogate Model of Chemical Flooding Based on the Artificial Neural Network for Enhanced Oil Recovery Implications," Mathematical Problems in Engineering, vol. 2015, Article ID 706897, 9 pages, 2015. doi:10.1155/2015/706897.

CHAPTER 2

Zackrisson, M. , Jönsson, C. and Olsson, E. (2014) Life Cycle Assessment and Life Cycle Cost of Waste Management—Plastic Cable Waste. Advances in Chemical Engineering and Science, 4, 221-232. doi: 10.4236/aces.2014.42025.

CHAPTER 3

J. Luis Gomes Marinho, R. Swarnakar, S. Rodrigues de Farias Neto and A. Gilson Barbosa de Lima, "Unsteady Fluidynamic Behavior of Gas Bubbles Flowing in Curved Pipes: A Numerical Study," Advances in Chemical Engineering and Science, Vol. 2 No. 2, 2012, pp. 283-291. Doi: 10.4236/aces.2012.22033.

CHAPTER 4

Michael Nikolaou, Computer-aided process engineering in oil and gas production, Computers & Chemical Engineering, Volume 51, 5 April 2013, Pages 96-101, ISSN 0098-1354, http://dx.doi.org/10.1016/j.compchemeng.2012.08.014.

CHAPTER 5

Edna Cabecinha, Martinho Lourenço, João Paulo Moura, Miguel Ângelo Pardal, João Alexandre Cabral, A multi-scale approach to modelling spatial and dynamic ecological patterns for reservoir›s water quality management, Ecological Modelling, Volume 220, Issue 19, 10 October 2009, Pages 2559-2569, ISSN 0304-3800, http://dx.doi.org/10.1016/j.ecolmodel.2009.06.011.

CHAPTER 6

Dai H, Mao J, Jiang D, Wang L (2013) Longitudinal Hydrodynamic Characteristics in Reservoir Tributary Embayments and Effects on Algal Blooms. PLoS ONE 8(7): e68186. doi:10.1371/journal.pone.0068186.

CHAPTER 7

Zhenyao Shen ,Lei Chen,Qian Hong Hui Xie,Jiali Qiu,Ruimin Liu,Vertical Variation Of Nonpoint Source Pollutants In The Three Gorges Reservoir Region, doi:10.1371/journal.pone.0071194.

CHAPTER 8

Kaushik R, Balasubramanian R, Dunstan H (2014) Microbial Quality and Phylogenetic Diversity of Fresh Rainwater and Tropical Freshwater Reservoir. PLoS ONE 9(6): e100737. doi:10.1371/journal.pone.0100737.

CHAPTER 9

Liu Q, Liu S, Zhao H, Deng L, Wang C, et al. (2013) Longitudinal Variability of Phosphorus Fractions in Sediments of a Canyon Reservoir Due to Cascade Dam Construction: A Case Study in Lancang River, China. PLoS ONE 8(12): e83329. doi:10.1371/journal.pone.0083329.

CHAPTER 10

T. Wilke, M. Schneider and K. Kleinermanns, "1, 4-Hydroquinone is a Hydrogen Reservoir for Fuel Cells and Recyclable via Photocatalytic Water Splitting," Open Journal of Physical Chemistry, Vol. 3 No. 2, 2013, pp. 97-102. doi: 10.4236/ojpc.2013.32012.

Index

A

Agricultural Nonpoint Source pollution (AGNPS) 157
Air injection time 56, 58, 62
Alkaline/surfactant/polymer (ASP) 2
Artificial neural network (ANN) 2, 4

B

Bio-available phosphorus (BAP) 210
Bottom hole pressure (BHP) 77, 80

C

Cellular Automata (CA) 92, 94, 106, 107, 109
Colony forming units (CFU) 182
Competitive algorithm (ICA) 7
Computer modelling group (CMG) 18
Corine Land Cover (CLC) 101

D

Danish Hydraulic Institute (DHI) 153
Digital Elevation Model (DEM) 157
Direct methanol fuel cell (DMFC) 229, 231, 240, 242

E

Enhanced oil recovery (EOR) 18